江苏省高等学校重点教材(编号:2021-2-015)
南京信息工程大学教材建设基金资助项目

Laboratory Experiments for Organic Chemistry

有机化学实验

主　编　李　英　邵　莺
副主编　Elseddik Abdelkader　焦　岩

特配电子资源

- 配套课件
- 视频学习
- 拓展阅读

南京大学出版社

图书在版编目(CIP)数据

有机化学实验 = Laboratory Experiments for Organic Chemistry /李英，邵莺主编. — 南京：南京大学出版社，2021.11
ISBN 978-7-305-24358-5

Ⅰ. ①有… Ⅱ. ①李… ②邵… Ⅲ. ①有机化学—化学实验—高等学校—教材 Ⅳ. ①O62—33

中国版本图书馆 CIP 数据核字(2021)第 060502 号

出版发行　南京大学出版社
社　　址　南京市汉口路 22 号　　　　邮　编　210093
出 版 人　金鑫荣

书　　名　有机化学实验
Laboratory Experiments for Organic Chemistry
主　　编　李　英　邵　莺
责任编辑　刘　飞　　　　　　　编辑热线　025-83592146

照　　排　南京南琳图文制作有限公司
印　　刷　广东虎彩云印刷有限公司
开　　本　787×1092　1/16　印张 16　字数 370 千
版　　次　2021 年 11 月第 1 版　2021 年 11 月第 1 次印刷
ISBN 978-7-305-24358-5
定　　价　49.00 元

网址：http://www.njupco.com
官方微博：http://weibo.com/njupco
官方微信号：njuyuexue
销售咨询热线：(025) 83594756

Preface

It came to our awareness that most *Sino*-Foreign cooperative undergraduate programmes in chemistry or related disciplines have limited options in practical chemistry textbooks that are in English. Either the textbook is imported and very expensive or limited practical textbooks are locally published. On the other hand, most instructors in chemical practical courses may be forced to choose experiments from several popular and classic laboratory textbooks that are written in Chinese. We have undertaken the organic chemistry practical education in English in the Applied Chemistry programme cooperated between the University of Reading and Nanjing University of Information Science & Technology (NUIST) since 2016. We have constantly modified our experiments and imported the experimental projects from the department of chemistry in the University of Reading. In the spring of 2018, our team agreed to the importance of formally publishing an organic chemistry laboratory textbook, which must include all the organic projects carried out in the cooperative undergraduate programme in NUIST. Meanwhile, we invited Dr. Ying Shao from Changzhou University to join our team and incorporated the organic chemistry experimental projects that are adopted in the foreign student programme in chemistry or related majors in Changzhou University.

We tried our best to emphasize extensively on the safety issues in the textbook as this is the first priority of any experiment. Various fundamental organic skills and techniques are introduced in the sections of "Carrying out an experiment" and "Basic techniques in organic chemistry" and most of them are involved in the detailed experimental part. Characterization techniques, such as the modern spectroscopic technologies, are briefly discussed. For each detailed experiment, objectives, list of hazards and equipment, principle of the experiment, detailed procedures, technique checklist, pre-lab discussion, experimental outline, helpful hints, and expected results are provided. Instructors and students are expected to follow the experiment easily.

It would be impossible to pull all together in this book without the commitment and hard work of all authors, who studied or worked in the organic chemistry field in Europe or North America. The valued experience in teaching organic chemistry practical courses helps to improve this book. Hundreds of students graduating from the programmes were involved in the validation, reproduction and input/comment to improve these experiments and are acknowledged herein.

We would also like to thank Dr. Bin Xue and Dr. Sheng Yin from the School of Physics, Nanjing University for preparing some key figures and graphs for the textbook.

We recognize that imperfectness and errors cannot be avoided. Therefore, we welcome and value any comments and criticism to this textbook.

Ying Li

2021.11

前　言

目前，我国高校不断扩大本科留学生的招生规模，同时进一步与国外高校开展中外合作办学并开设相关专业。另外，很多高校为提高专业人才的综合素质和质量，也在积极尝试开设双语，甚至全英文的专业课程。“有机化学实验”是高校化学、化工、材料、医药等专业普遍开设的一门基础实验课程，双语或全英文授课对学生未来发展的培养优势明显。

南京信息工程大学与英国雷丁大学(NUIST-UoR)合作办学开设的应用化学专业，已在2016年进入第二阶段，项目要求理论与实验的实际教学均通过全英文开展，而目前国内适用于“有机化学实验”英文课程的教材较为稀少，进口原版教材价格昂贵，而且不能很好地兼容我国高校有机化学实验教学体系。

本书是依据南京信息工程大学应用化学专业(中外合作)和常州大学化学专业(留学生)有机化学实验项目的基础，综合部分我国高等院校有机化学实验常规内容，所编写的全英文实验教材。本书不仅保留了国内传统有机化学实验教学的经典项目，还引入了英国雷丁大学先进的实验教学内容。比如，本书在充分强调实验室安全的基础上，首次引入了英国危险健康控制体系(Control of Substance Hazardous to Health)，通过定性与定量评估化学品及操作的安全性与应对措施的策划，进一步确保实验安全理念能植入师生脑中。同时，书中在“Carrying out an Experiment”和“Basic Techniques in Organic Chemistry”两个部分系统地介绍了一系列的基础有机实验技术与技能；在“Application of Spectroscopy in Organic Laboratory”中介绍了利用光谱学方法表征有机化合物的基本原理和操作方法；在“Organic Experiments”部分囊括了近35个从常规到具有探索性的有机化学实验，涵盖了基本的有机反应类型和有机实验单元操作。本书每个实验项目均包括实验目的、危险品与仪器清单、实验原理、实验步骤、技术清单、预实验讨论和预期结果等模块。我们期望，有机化学实验的师生都能很容易地使用本书，学生更能在实验操作与总结报告撰写过程中学习有机化学领域的科学研究过程。

本书既可作为高校与化学相关专业的双语、全英文课程教材，也可作为国际合作办学项目的课程用书。

本书主编为南京信息工程大学李英教授和常州大学邵莺教授，全书的编者均具有欧洲、北美的留学或工作经历，并且结合自身多年的实验教学经历，投入了大量的时间和精力编写本书。此外,上百名历届 NUIST-UoR 应用化学中外合作办学项目的学生们对课程教学提出的良好建议,也对本书中各个实验项目的重复与验证的完善与提高提供了帮助。最后,南京大学物理学院薛斌博士、尹晟博士对书中部分图片进行了设计与完善。在此,一并表示衷心的感谢。

本书编撰过程中,我们力求完善,但由于编者水平和经验有限,错漏在所难免,诚望各位同行和读者提出批评与宝贵建议,以便在后续修订时进一步完善提高。

李　英

2021 年 11 月

Content

1 Introduction

1.1 Safety in the Laboratory

Personnel safety and laboratory safety are always the key priority and prerequisite for any lab work. However, there is a high potential that one may get hurt in a laboratory if no attention has been paid to safety. Therefore, everyone in the laboratory is responsible for the safety of the laboratory so that everyone in the laboratory can work safely, smartly, and happily.

When entering the laboratory, one should first check the locations and functionality of the following safety equipment:

1. Fire extinguishers.
2. Fire blankets.
3. First-aid kits.
4. Safety showers and eye wash stations/face sprays.

Emergency calls, **110** or **119** as in China, are available for anyone who needs help or is in critical dangers.

Briefly, only CO_2 and dry-chemical fire extinguishers should be used (after training) on chemical or electrical fires. Water faucets at sinks may be used to wash skin exposed to corrosive chemicals for at least 15 minutes. One should check the location of safety equipment in the working areas when first entering a laboratory and be sure (even rehearse) what to do in the case of a fire or other accidents. However, in the event of fire or other accidents, do not take any action that would risk the safety of oneself or others. Most importantly, make any emergency notified as soon as possible to a staff or the instructor.

Anyone in the laboratory is mandatorily required to wear a **lab coat** and a pair of **safety goggles** at all times. *This is a general lab safety law in various research institutes*. Gloves are usually provided in the undergraduate laboratory for students to handle chemicals. The choice of gloves can be found in Table 1. 1. Although radios and musical instruments are not technically considered as safety hazards, they will not be allowed in an undergraduate laboratory.

Table 1.1 Property of commonly used gloves to various compounds

Compound	Glove type		
	Neoprene	Nitrile	Latex
Acetone	√√	√	√√
$CHCl_3$	√√	×	×
CH_2Cl_2	√	×	×
Et_2O	√√√	√√	×
EtOH	√√√	√√√√	√√√√
EtOAc	√√	×	√
Hexane	√√√√	√√√√	×
MeOH	√√√	√	√
HNO_3 (conc.)	√√	×	×
NaOH	√√√	√√√√	√√√√
H_2SO_4 (conc.)	√√	×	×
PhMe	√	√	×

Note: × poor, √ fair, √√ good, √√√ very good, √√√√ excellent

Learning about the hazards of materials, equipment, and procedures used in chemical laboratories is a part of the educational objective. One must learn the hazards of chemicals that might be encountered during the lab hours before entering the lab.

Disposal of solvents, chemicals and other materials:

Never pour solvents or reactive chemicals down the drain or throw chemicals into regular garbage cans. Such careless handling of flammable or toxic chemicals presents a serious hazard in the laboratory and raises risks to the environment. Also, never keep an open beaker of such solvents/reagents outside a fumehood. They have to be collected in special solvent waste containers or solid waste containers provided in the lab. When in doubt about how to dispose of something, ask the instructor. If drain disposal is necessary and acceptable, always flush the drain before, during, and afterwards with a lot of water. **All broken glass must be discarded in specially designed containers (called glass box) and one must report such event.** A dustpan and brush for broken glass can be checked in the lab. Spilled mercury is a special safety hazard and should be reported to the instructor for cleanup.

General Safety Rules for An Undergraduate Laboratory

a. The safe way is the right way to do one's job. Plan your work ahead of time and follow instructions. If you do not know how to do the experiment safely, check with the instructor.

b. Be able to use all safety devices and protective equipment provided in the laboratory and know their locations (eyewash fountain, safety shower, fire blanket, fire extinguisher, and first-aid kits).

c. Safety goggles are mandatory to be worn at all times.

d. Appropriate gloves must be worn when handling chemicals. The properties of some commonly used gloves are summarized in Table 1. 1.

e. Do not eat or drink in the laboratory. Smoking in the laboratory is absolutely forbidden.

f. Proper clothing should be worn (including protective clothing when handling corrosive, toxic, or flammable materials). Avoid loose sleeves, loose cuffs, bracelets, sandals and open-toes. Be careful with long hair.

g. Horseplay in any form is dangerous and prohibited. Do not run in laboratory areas.

h. If you see a colleague doing something dangerous, point it out to him or her and report it to the instructor.

i. Report to your instructor all unsafe conditions, unsafe acts, and "near-misses" that might cause future accidents. Report any accident or fire, no matter how trivial, to the instructor.

j. Hazardous Chemicals:

   a) Be especially mindful of fire hazards when you or your lab neighbors are working with flammable liquids.

   b) Know common explosive, toxic, and carcinogenic materials and use them only with adequate safeguards.

   c) Assume any chemical is hazardous and handle it carefully with full protection in the hood if you have no information about the chemical.

k. Never leave a reaction or experiment running unattended, unless you have told your lab partners enough about it to deal with potential hazards while you are away.

l. Keep hood and bench top areas clean and workable space maximized.

1.2 Chemical Hazards

Any chemical has specific physical and chemical properties. Precautions must be taken to transport and handle any chemical in order to prevent potential health and environment risks. The United Nations has set up a whole standard classification and labelling system of chemicals, which is known as the *Globally Harmonized System of Classification and Labelling of Chemicals*, short as GHS. The GHS includes standardized criteria to test hazards, universal warning pictograms, and harmonized material safety data sheets (MSDS), recently renamed as just SDS (safety data sheets).[①] According to GHS system, hazards are classified into three classes: Physical hazards, health hazards, and environmental hazards. The GHS physical hazards provide qualitative and/or semi-qualitative criteria with multiple hazard levels. It is necessary to identify physical states of substances for the physical hazard criteria. A gaseous substance should have a vapor pressure higher than 0. 3 MPa at 50 ℃ or is in gas state at 20 ℃ and at a standard atmospheric pressure of 0. 101 3 MPa. A liquid is a non-gaseous compound or mixture which has a melting point no higher than 20 ℃ at a standard atmospheric pressure of 0. 101 3 MPa. A solid is a substance or mixture which is not in its gas phase or liquid phase at 20 ℃ and at a standard atmospheric pressure of 0. 101 3 MPa. Based on the chemical properties of a substance, the physical hazards can be briefly described below.

- Explosive——a compound is capable of producing gas via a chemical reaction in a rapid fashion and causing severe damage to the surroundings due to burst pressure produced.
- Flammable gas——a gas has a flammable range in air at 20 ℃ and a standard atmospheric pressure of 0. 1013 MPa.
- Flammable liquid——a liquid has a flash point of no higher than 93 ℃.
- Flammable solid——a solid is combustible easily or cause a fire via friction.
- Flammable aerosol——an aerosol compound contains any component regarded as flammable based on the GHS criteria.
- Oxidizing gas——a gas causes or contributes to the combustion of other substances by providing oxygen more efficiently than air.
- Oxidizing liquid——a liquid causes or contributes to the combustion of other substances by yielding oxygen while itself may not be combustible.
- Oxidizing solid——a solid causes or contributes to the combustion of other substances by yielding oxygen while it alone may not be combustible.
- Gas under pressure——a gas in a container has a pressure no less than 200 kPa at 20 ℃ or is stored as a refrigerated liquid.
- Self-reactive substance——a compound is thermally unstable and undergoes a rapid and strongly exothermic decomposition even in the absence of oxygen.
- Pyrophoric liquid——a liquid is readily to ignite within 5 minutes in contact with air.
- Pyrophoric solid——a solid is readily to ignite within 5 minutes in contact with air.

---

① *A Guide to The Globally Harmonized System of Classification and Labeling of Chemicals* (GHS). 1st ed., United Unions, 2003.

- Self-heating substance——a non-pyrophoric compound is prone to self-heat in air without energy input.
- Substance which in contact with water emits flammable gases——a non-gaseous substance emits spontaneously flammable gases or gives a dangerous quantity of flammable gases in contact with water.
- Organic peroxide——an organic compound with the structural feature of R—O—O—R′.

Health hazards are the ones concerning the potential short-term or long-term hazardous effects on health, including acute toxicity, skin corrosion/irritation, eye irritation/damage, respiratory or skin sensitization, germ cell mutagenicity, carcinogenicity, reproductive toxicology, target organ systemic toxicity of single exposure or repeated exposure, and aspiration toxicity. Environmental hazards are classified as chemicals that are hazardous to the aquatic environment. They include acute aquatic toxicity and chronic aquatic toxicity. The chronic aquatic toxicity refers to either bioaccumulation potential or rapid degradability.

Table 1.2 The universal pictograms of chemical hazards

Symbol	Hazardous	Symbol	Hazardous	Symbol	Hazardous
	Carcinogen, respiratory sensitizer, reproductive toxicity, target organ toxicity, mutagenicity, aspiration toxicity		Explosive, self-reactive, organic peroxides		Irritant, dermal sensitizer, acute toxicity (harmful), narcotic effect, respiratory tract irritation
	Corrosive		Flammable, self-reactive, pyrophoric, self-heating, emitting flammable gas, organic peroxides		Oxidizing
	Environmental toxicity		Gases under pressure		Toxic, fatal

The most used way to describe the hazard is the pictograms shown in Table 1.2. A chemical hazard might belong to one or more types as listed in Table 1.2. These symbols can be found on the label of a chemical container to warn that special precautions and disposal methods are required. To identify the specific type(s) of a hazard, the SDS of this hazard should be obtained. SDS is available in various online sources. A good and recommended source for the hazard's safety information is the website[①] of in the Merck KGaA. Although Sigma-Aldrich is an international chemical, life science, and biotechnology company, it provides relatively detailed information on "Safety & Documentation" "Protocols & References" "Ratings & Reviews" in addition to the purchase information such as catalogue number, availability, and price. For each chemical, codes for hazard statements and corresponding precautionary statements can be found. The codes for hazard statement can specify and detail the hazardous information. For instance, there are 5 hazard statements for diethyl ether (CAS No. 60 – 29 – 7) and correspondingly 4 precautionary statements as shown in Table 1.3. Each statement has a specific code.

Table 1.3 The hazardous information of diethyl ether

Code	Hazard statement	P-Code	Precautionary statement
H224	Extremely flammable	P210	Avoid heat/sparks/flames/hot surfaces
H302	Harmful if swallowed	P240	Ground/bond container and receiving equipment
H336	Potential effect of drowsiness or dizziness	P304+P340	Remove the victim to fresh air and keep at rest in a position comfortable for breathing if inhaled
EUH019	Potential formation of explosive peroxides	P403+P233	Store in a well-ventilated place. Keep container tightly closed
EUH066	Potential effects on skin dryness or cracking for long-term and repeated exposure		

Note: EU is the supplemental hazard information.

Based on the extent the hazard class has, the chemical's hazard can be further classified with different hazard categories as summarized in Table 1.4. For instance, diethyl ether can be classified in category Ⅳ because it has acute toxicity if orally taken, classified in category Ⅲ because it is a flammable liquid, and classified in category Ⅰ because it raises specific target organ toxicity of single exposure.

① http://www.sigmaaldrich.cn/.

Table 1.4 List of hazard categories

Hazard category		Detailed properties describing the hazard	Examples
Ⅳ	Low	Of mild or no toxicity, corrosion, flammability, or harm	Water, N_2, CO_2, glucose
Ⅲ	Medium	Toxic, corrosive, oxidizing, harmful, superior flammable	CO, EtOH, NaOH, MeOH, pyridine
Ⅱ	High	Very toxic or any combination of two among "toxic" "corrosive", and "oxidizing"	Aniline, hydrazine, nicotine, ozone
Ⅰ	Extreme	Known or suspected to be carcinogenic, or extremely toxic	Benzene, HCN, mercury, tetraethyl lead

1.3 Use of a COSHH Form

COSHH is short for the **Co**ntrol of **S**ubstance **H**azardous to **H**ealth. COSHH includes the regulations and laws in the United Kingdom that helps employers to control chemicals that may be hazardous to human health and may raise environmental issues. It can help to assess the safety of work, prevent or reduce chemical exposure of people (workers, researchers, etc.), and provide precautions and incident planning. These regulations have been effective for more than twenty-five years in the United Kingdom and can be introduced to assess the hazards safety and precautions for laboratory work. Typically, COSHH forms would help to achieve the following:

1. To identify the health hazards.
2. To plan ahead of time in order to prevent the hazardous effects to human health and environment.
3. To provide a control procedure to lower any potential harm to human health.
4. To ensure the usage of the chemicals is appropriate.
5. To provide procedures, instructions, and training for people dealing with the hazards.
6. To provide close monitoring and health surveillance in the situations when chemicals are used.
7. To provide suitable information in emergencies.

Although the experimental procedure is pivotal, there is nearly no information regarding the hazards and safety in the procedure description. In the procedure, the quantities of the chemicals should be indicated and the way to use them may be detailed. However, it is important for anyone who is going to carry out the experiment to know the potential harm and precaution of using each chemical.

The COSHH form is used to evaluate the hazards we may work with. Figure 1.1

is an empty COSHH form used in an undergraduate chemistry laboratory. In this form, a brief experimental procedure should be described and the risk of using chemicals should be assessed by stating the amount of each chemical used in the experiment, its hazard description, and its risk code. The hazard description and risk code are available in the SDS of each chemical. Precautions and control procedures should be provided to minimize all possible safety issues. As waste is usually generated, the way to dispose the waste should also be clarified, especially for the extremely dangerous hazardous chemicals.

Undergraduate Risk Assessment Form

Experiment				Course	
Student				Date	
Supervisor		Highest Risk Code		Fume cupboard	
				Gloves	

Outline of experimental procedure

Substance	Quantity	Hazard description Corrosive Irritant etc.	COSHH Risk code

Precautions and control measures

Waste Disposal

Undergraduate Risk Assessment Form
Data For High Hazard category chemicals when >1 g and used in an open system or >100 g

Substance	Quantity	Hazard Category	Activity	Exposure Scores A	B	C	total	Exposure potential	Risk Code
Ethanol	<33 mL	MED	Dispensing(open)	10	10	1	100	LOW	LOW

Additional Notes
N/A

Figure 1.1 An empty Baum COSHH form

Table 1.5 Evaluation for the exposure score

Category		Exposure score		
		1	10	100
A	Quantity	<1 g	1~100 g	> 100 g
B	Physical characteristics	Non-volatile liquid, dense solid, of no skin adsorption	Volatile liquid, lyophilized solid, dusty solid, of low skin adsorption	Gases, highly volatile liquid, aerosols, of high skin adsorption
C	Characteristics of activity	Mainly within an enclosed system and of low chance of accident	Partially open system and low chance of accident	No physical barrier or medium to high chance of accident

In a COSHH form, the exposure score for a substance must be identified to estimate the exposure potential. There are three categories to be scored: Quantity, physical characteristics, and characteristics of the activity. The score of each category relies on the individual detail. As demonstrated in Table 1. 5, use of higher quantity of a chemical would obviously raise higher exposure and hence the exposure score is correspondingly higher. In addition, handling a solid is usually much safer and more controllable than handling a gaseous compound because gaseous compounds would have a higher potential of contact with a larger surface area of the human-skin. Moreover, it is always recommended to work inside a fumehood to maximize the protection of the operator. Once the exposure scores for each category is obtained (A, B, or C), the overall exposure score of handling a certain compound is defined as the product of A, B, and C as demonstrated in equation (1. 1). The exposure potential of such compound in a procedure is determined by its exposure score based on Table 1. 6.

$$Exposure\ score = A \times B \times C \qquad \text{equation (1. 1)}$$

Table 1. 6 Determination of exposure potential

Overall exposure scores	<1 000	1 000~10 000	>10 000
Exposure potential	Low	Medium	High

To further obtain the risk code of a hazard, both the hazard category and the exposure potential must be considered at the same time. The former mainly describes the hazardous properties of the chemical as described in the Chemical Hazards section, while the latter concerns the detailed information on how the chemical is handled. By using a matrix table, it would be easy to identify the risk code of a hazard as shown in Table 1. 7.

Table 1. 7 The matrix for the risk code of a hazard

Hazard category		**Risk code for the exposure potential**		
		Low	**Medium**	**Extreme**
Ⅰ	**Extreme**	Extreme risk (E)	Extreme risk (E)	Extreme risk (E)
Ⅱ	**High**	Medium risk (M)	High risk (H)	High risk (H)
Ⅲ	**Medium**	Low risk (L)	Medium risk (M)	Medium risk (M)
Ⅵ	**Low**	Low risk (L)	Low risk (L)	Low risk (L)

When the risk of using a chemical is determined, precautions and control methods must be provided in addition to a waste disposal plan in the form.

A sample COSHH form that is used for an experiment is provided in section **1. 6.**

1.4 Pre-lab Reports or Notebooks Used in the Laboratory

It is important to prepare each laboratory experiment ahead of time and understand the corresponding principles and procedures before working in the laboratory. Instead of reading the content freshly in the wet lab, it is necessary for students to complete their preparation of each experiment by studying the experiment content, abstracting the key techniques involved, investigating the physical properties and safety information of the reagents used in the experiment, and clarifying the procedure of the experiment. A pre-lab report is useful to achieve the purpose, which does not equal to a direct copy of the protocol described in each experiment item of a laboratory manual or textbook. The students need to become clear on what to be done and what to be recorded during a lab period and data-recording spaces should be provided in the pre-lab report. During each wet laboratory, when the experiment is carried out in a laboratory, every student must use a pen to record the phenomena (observations) on the pre-lab report for each step, such as color changes, precipitation formation, bubbling, and shapes and colors of crystals. The actual operation in the laboratory and the key data, such as melting points, boiling points, weight of the crude product, and image of a developed thin layer chromatographic plate (for R_f calculation), should also be recorded.

A pre-lab report should contain a brief introduction and objective of the principle for the experiment, physical properties of reagents used in the experiment including boiling point, melting point, density of a liquid chemical, solubility in a certain solvent, *etc.*, simple but easy to follow procedure, and data recording section. A flow chart or step-wise description can be used for the experiment procedure. The data recording section might be an individual section or merged in the procedure.

General guidelines for the pre-lab report:

1. Start a new page for each new experiment.

2. Write the title of the experiment, date, and your name at the top of each page.

3. Indicate if a page is continued from the previous page.

4. Never skip a space for later additions.

5. Be neat and thorough! Someone should be able to pick up your notebook twenty years from now and be able to repeat your experiments.

6. List the physical properties and amounts of chemicals used in the experiment.

7. A sketch of the assembly of glassware for the reaction and operation is

recommended to be provided.

Recording during laboratory hours:

1. Use a pen not a pencil to record your data.

2. Do not erase or use whiteout. If you make a mistake, draw a single line through the error and write the correct entry on the top or side of it.

3. Do not remove an original page. If the entire page is incorrect, draw a single diagonal line through the page and sign the name and date.

4. Record all phenomena and data (with units) directly into your notebook during the experiment.

(1) DO NOT record data on scrap paper, your hand, *etc.*, to be transferred later.

(2) If you need to enter remote data (printout from computer, etc.), date, sign, and tape it into your notebook.

(3) Record the actual operation and reagents' amounts if they are different from the pre-lab report.

(4) Record the assembly of glassware if that is different from what is drawn in the pre-Lab report.

1.5 Lab Reports

Post the laboratory, students are expected to summarize the actual protocol, which can be different from what is described in the manual, analyze the phenomenon and data, discuss the experiment based on the phenomena and data in order to provide reasonable explanation, give meaningful rationales to evaluate the failure or success of the experiment, draw a final conclusion on the experiment, and provide reasonable advices to improve the experiment. Therefore, a lab report is usually composed of *Abstract*, *Introduction*, *Experimental*, *Analysis* and *Discussion*, *Conclusion*, and *Reference*.

- *Abstract* is used to summarize the method, findings and conclusion of a project. It should be provided in the form that the audience can quickly capture the core content of the project and may become interested in the detailed information.

- *Introduction* usually functions as a background summarization of the related field or describes the key principle and purpose of the project. It may also state the reason to choose the specific methods. Usually, at the end of the *Introduction*, a brief summary of the work is provided.

- *Experimental*, or sometimes *Materials and Methods*, is the part to organize and summarize the detailed procedures, phenomenon, weight, and yields. The procedure reported in the lab report might be different from the pre-lab report as it is subject to changes. The students should summarize this part the same as what they have done exactly in the laboratory. Characterization results, such as boiling points, melting points, and R_f, should be reported in this section if applicable. For synthetic experiments, the yield of the reaction should be reported as it is typically a quick and direct evaluation of the experiment using the following equation(1. 2):

$$reaction\ yield = \frac{m_{\text{obtained}}}{m_{\text{theoretical}}} \times 100\% \qquad \text{equation(1. 2)}$$

Where m_{obtained} is the actual mass of the isolated product while $m_{\text{theoretical}}$ is the maximum amount of the product that can be obtained if all limiting reagent is converted completely into the product.

• *Analysis and Discussion*, or sometimes *Results and Discussion*, is the core part of a lab report. The students should try to provide a reasonable analysis and explanation to the results in additional to various phenomena. This can provide a clue to the success or failure of the experiment, and may also help to discuss a better and potential improvement of the experimental design. Based on one's personal experience to conduct the organic experiment, one may use this section to reflect and perfect one's operation if similar experiment or operation will be used in the near future.

• *Conclusion* is to state the overall findings and main points for the experiment, which should be neater than the *Analysis and Discussion* section.

A sample lab report is provided in the following section.

1.6 Samples of Reports and COSHH Form

In this section, samples of the documents required for a complete undergraduate organic laboratory are provided. Students need to finish the COSHH form and pre-lab report before entering the laboratory, record phenomena and data during the lab hours, and organize all information and complete the lab-report post the laboratory. It is highly recommended that students follow the formatting of each file. It must be noted that a lab report is never a pre-lab report plus experimental data. The lab report is a summary of what a student has actually done and achieved in addition to data analysis, discussion, and conclusion.

Here the first experiment in this book is used as a sample. The grey part is the recording or writing before or during the laboratory and usually handwriting for the phenomena and data is applied.

A. A sample pre-lab report

Pre-lab Report

Purification of Liquids by Distillation

Sharon Cao
200018001219
The NUIST Reading Academy
Nanjing University of Information, Science & Technology

Purpose

The purpose of this experiment is to apply atmospheric pressure distillation to isolate ethanol from water in addition to determine the concentration of the unknown aqueous ethanol solution.

Principle

Atmospheric distillation makes use of the intrinsic different boiling points of different liquid chemicals to purify or isolate the components in a mixture consisted of these liquid chemicals by the phase transitions between gas state and liquid state under the atmospheric pressure. To apply distillation separation, it is necessary for the liquid components to have a relatively large boiling point gap with a minimum of 20 ℃ difference. However, it should be clear that some components can form azeotropes and these components will be co-distillated out with a certain composition at the boiling point, which is distinct from and always lower than that of each component. For instance, ethanol and water can form an azeotrope containing 4.5% water at 78.2 ℃ under the atmospheric pressure. Via atmospheric distillation, only 95.5% ethanol in water can be obtained. This can be used to determine the concentration of the initial aqueous ethanol solution if all low boiling fraction is collected and weighed by using the following formula:

$$x=\frac{m_o \times 95.5\%}{m_i} \times 100\%$$

in which m_i is the weight of the initial aqueous ethanol solution, and m_o is the weight of the collected fraction.

Materials and Methods

The aqueous ethanol solution is provided by the laboratory. Ethanol is a transparent liquid with a boiling point of 78.4 ℃ under 1 atmosphere pressure. Boiling chips should be used to facilitate the boiling of the liquid.

Some glassware and equipment are used: A 50 mL round bottom flask (RBF), a distillation head, a thermometer, a thermometer adapter, a Liebig condenser, water tubing, a receiving adapter, a 50 mL Erlenmeyer flask, two iron stands, two clamps and two clamp holders. The joints of glassware should fit appropriately to avoid leakage of vapor or liquid. A heating source must be provided to facilitate the boiling/evaporation of the liquid.

The procedure of this experiment is summarized in two ways: A flow chart and a step-wise description. (You can choose one over the other.)

A flow chart procedure

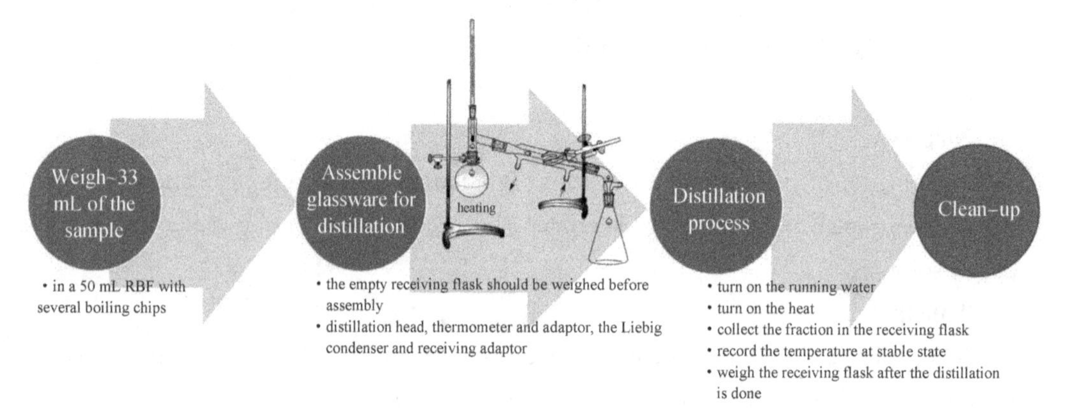

A step-wise description

i. Weigh 33 mL of the aqueous ethanol solution in a 50 mL RBF containing several boiling chips and weigh the empty receiving flask (RF).

ii. Assemble the glassware as illustrated on the right with all joints connected firmly.

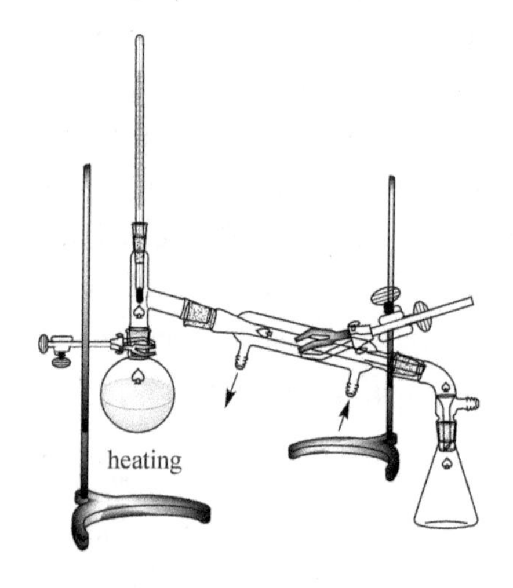

iii. Turn on the running water slowly.

iv. Turn on the power for the heating mantle to heat the RBF.

v. Record the boiling point when the thermometer reaches equilibrium.

vi. When the temperature drops dramatically, turn off the heat and remove the heating mantle to allow the RBF to cool down.

vii. Remove the RF and weigh it again.

viii. Submit the received fraction.

ix. Wash all glassware and clean up the bench.

Data recording

<table>
<tr><td colspan="2">Phenomena/observations during the procedure</td><td>Upon heating, some bubbles formed and the volume of the liquid in the RBF decreased. There were transparent liquid drops formed at the thermometer's glass ball. The reading of the thermometer increased quickly and vibrated between 75.1—77.5 ℃, during which there was always a drop hanging over the glass ball of the thermometer. Drops of colorless transparent liquid were collected in the RF. The liquid in the RF has some special odor.</td></tr>
<tr><td rowspan="3">Mass</td><td>33 mL of initial sample</td><td>29.2 g/31 mL, Sample **A**</td></tr>
<tr><td>Empty RF</td><td>86.3 g, a 100 mL round bottom flask was used instead of a 50 mL one</td></tr>
<tr><td>RF containing distillate</td><td>101.5 g</td></tr>
<tr><td colspan="2">Boiling point at a stable reading</td><td>75.1~77.5 ℃</td></tr>
</table>

B. A sample lab report

Lab Report

Purification of Liquids by Distillation

Sharon Cao

200018001219

The NUIST Reading Academy

Nanjing University of Information, Science and Technology

Abstract

We tried to use atmospheric distillation to purify ethanol azeotropically from aqueous ethanol solution and found that the ethanol concentration of Sample **A** was 49.7% and the boiling point of the collected fraction was 75.1~77.5 ℃.

Introduction

Atmospheric distillation is an important technique in chemical laboratory and industry. It has been widely used to purify liquid compounds over a liquid mixture. For example, the traditional Chinese winery workshops usually use atmospheric distillation to separate liquid components including alcohol and esters from the crude fermentation mixture to guarantee a better taste of the liquor. In addition to atmospheric distillation, other distillation techniques, including vacuum distillation and fractional distillation, are capable of purifying liquid compounds with high boiling points or of thermo-sensitivity and/or more complicated liquid mixtures. The key principle of distillation is based on the phase transition: First from liquid to gaseous phases and then from gaseous to liquid states, which is useful to isolate compounds of different boiling points easily. Some compounds may have interactions with each other either in the liquid phase or in the gaseous phase, and this may lead to azeotrope formation. Even though azeotropic distillation is not a complete process to separate one from the other but it is routinely used to concentrate some solutions or get rid of one component in a mixture, such as newly formed product in a reaction.

Ethanol is a common solvent and a useful chemical in organic chemistry. It is miscible to water. The boiling point of ethanol is 78.4 ℃ while that of water is 100 ℃ at one atmospheric pressure. The large gap in boiling points makes the distillation a perfect means to separate the two liquids. However, ethanol can also form azeotrope with water and boils at 78.1 ℃ with a vapor that contains 95.5% ethanol. Therefore, atmospheric distillation can be used to remove a large amount of water in aqueous ethanol solutions. Furthermore, as the azeotropic distillation only affords a mixture of a constant concentration of ethanol, it can be used to determine the concentration of the initial solution by measuring the masses of the initial solution and the collected distillate as shown in equation 1:

$$x\% = \frac{m_o \times 95.5\%}{m_i} \times 100 \qquad \text{equation 1}$$

in which m_i is the weight of the initial aqueous ethanol solution, and m_o is the weight of the collected fraction.

In this experiment, we tried to use the atmospheric distillation to determine the concentration of an aqueous ethanol solution **A**.

Materials and Methods

Sample **A**, as one of the aqueous ethanol solutions provided by the undergraduate laboratory was used for this experiment.

Around 31 mL of sample **A** was weighed to be 29.2 g in a 100 mL round bottom flask① containing several boiling chips. A distillation head, a thermometer adapter with a thermometer, a Liebig condenser with rubber tubing, a receiving adapter and a 50 mL pre-weighed Erlenmeyer flask were successively assembled tightly to the round bottom flask. After the running water was turned on, the round bottom flask was heated over a heating mantle at ~90 ℃ and bubbles formed during the heating process. It was observed that drops were formed at the glass ball of the thermometer and fell back to the round bottom flask. When the temperature dropped and no additional liquid was collected in the receiving flask, the heating source was removed. About 15.2 g of a colorless and transparent fraction in a temperature range of 75.1 ℃ to 77.5 ℃ was collected in the pre-weighed 50 mL Erlenmeyer flask and 52.1%

① Only 100 mL round bottom flask was available in the laboratory.

collection rate was achieved[①]. The collected liquid has a pleasant smell.

Analysis and Discussion

The formation of bubbles during the distillation is one of the evidences of the phase transition from liquid to gas upon heating. The reading of temperature for the fraction collected varied from 75.1 ℃ to 77.5 ℃, which was clearly somewhat lower than reported. This might be related to the lower atmospheric pressure in the laboratory. As the composition of the distillate is 95.5% ethanol and 4.5% water, the original concentration of sample **A** is calculated to be 49.7% by equation 1. This value might be slightly lower than the actual concentration, because there was some residual distillate in the condenser and the receiving adapter, which was difficult to be completely collected and weighed. Furthermore, for a liquid volume of ~31 mL, a container with a size of 50 mL is more appropriate than the 100 mL round bottom flask that was used in this experiment. Because in the large volume size of the container, there would be a relatively high amount of vapor remained to fill the space of the flask, which could contribute to the loss of azeotrope collected in the container and leads to a lower result than the actual concentration. It is also worth noting that the control of temperature is critical for the distillation. As for this experiment, we used the heating mantle as the heating source and we tried very hard to adjust its heating power to avoid over-heating of the distillation flask. It would be imagined that a strong heating would drive the evaporation of the residual water quickly, which would contribute to the mass of the collected fraction.

Conclusion

Overall, the experiment was a success. We have successfully used atmospheric distillation to collect ethanol-water azeotrope and determined the concentration of the provided sample to be 49.7%. The technique is useful in liquid separation.

Reference

Santosh Kumar, Neetu Singh, Ram Prasad. Anhydrous ethanol: A renewable source of energy, *Renewal and Sustainable Energy Reviews*, 2010, 14(7), 1830 - 1844.

① For a synthetic experiment, an isolated yield is usually reported instead of the collection rate.

C. A sample filled COSHH form

<table>
<tr><td>Experiment</td><td colspan="3">Purification of Liquids by Distillation</td><td>Course</td><td>Fundamental Chemistry Experiment</td></tr>
<tr><td>Student</td><td colspan="3">Sharon Cao</td><td>Date</td><td>Apr 5th, 2020</td></tr>
<tr><td rowspan="2">Supervisor</td><td rowspan="2">Dr. Shannon Li</td><td rowspan="2">Highest Risk Code</td><td rowspan="2">L</td><td>Fume cupboard</td><td>Yes</td></tr>
<tr><td>Gloves</td><td>Yes</td></tr>
</table>

Outline of experimental procedure
1. Weigh ~33 mL aqueous ethanol solution in a 50 mL round bottom flask and the receiving Erlenmeyer flask 2. Add several boiling chips to the RBF and assemble the distillation set-up 3. Turn on the running water and the Variac 4. Remove the Variac when the temperature drops and no additional drops at the same temperature range are collected 5. Weigh the receiving flask with the distillate

Substance	Quantity	Hazard description Corrosive Irritant etc.	COSHH Risk code
ethanol	<33 mL	Flammable	MED

Precautions and control measures
Open flame should be avoided. Eye protection and hand protection should be provided.

Waste Disposal
Ethanol solution should be collected in the non-halogenated solvent waste. The boiling chips are dumped to the solid waste. The water residue is flushed in the drain.

Data For High Hazard category chemicals when >1 g and used in an open system or >100 g

Substance	Quantity	Hazard Category	Activity	Exposure Scores				Exposure potential	Risk Code
				A	B	C	total		
Ethanol	<33 mL	MED	Dispensing (open)	10	10	1	100	LOW	LOW

Additional Notes

N/A

1.7 Commonly Used Laboratory Glassware

Many organic reactions and procedures involve various glassware. Some frequently used ones are listed herein. Briefly, the glassware can be classified on the basis of functions, such as containers, adapters, *etc.* The size or volume of the glassware is usually considered for the actual use in the organic laboratory. Some pieces of glassware have standard ground-glass joints, which are conically tapered joints and usually referred as standard taper glassware and symbolled with $××/××. For the standard taper glassware, the joints include the inner joints (or male joints) that have the outward side covered with ground glass surface, and the outer joints (or female joints) that have the inward side covered with ground glass surface, all of which are carefully ground with the same criteria to ensure that all the pieces can fit perfectly well and interchangeably. For instance, as shown in Figure 1. 2, an inner joint of a round-bottom flask might be labeled with $ 19/22 or S. T. 19/22, and it stands for that the outer diameter at the top of the joint, the wideness of the inner joint, is 19 mm, and the length of the ground glass for the joint is 22 mm. While an outer joint of a condenser might be labeled with $ 19/22 or S. T. 19/22 standing for the inner diameter at the top of the joint is 19 mm and the length of the ground glass for the joint is 22 mm. Therefore, the two pieces of glassware fit well with each other and can be chosen as a pair for a reaction requiring refluxing. The most used conically tapered joints are with the outer diameters of the inner joints or the inner diameters of the outer joints of 14 mm, 19 mm and/or 24 mm. The joint lengths vary because of different standards such as ASTM E676 - 02 and ISO 383. Accordingly, joints of No. 14, No. 19, and/or No. 24 are usually used for simplicity.

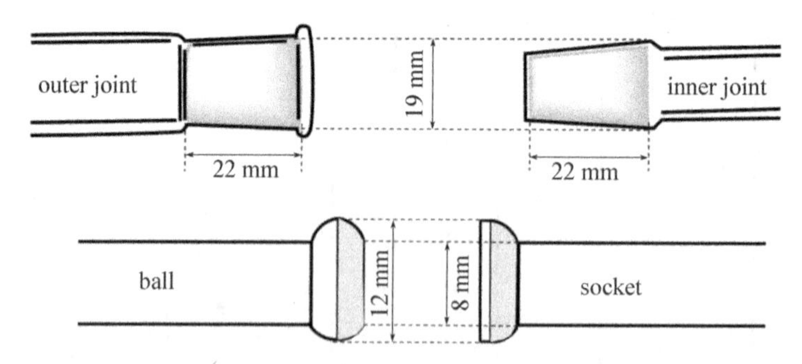

Figure 1. 2 An illustrative image of the taper standard joints $ 19/22 (top) and an illustrative image of ball-and-socket joint 12/8 (bottom)

In addition to the conically tapered joints, ball-and-socket joints, usually known as spheric joints, are used for some special glassware such as the solvent collecting system of a rotary evaporator and Schlenck lines. Such joints also have a pair of the inner joint and outer joint with ground glass and matching very well with each other. The inner joint is a ball while the outer joint is a socket, either of which holds a hole in the middle and located at the end of the corrsponding tube. The ball joint is a hemisphere and has the ground glass on the outside surface and the socket one has the ground glass coated on its inside surface to fit the ball tip perfectly. Different from the conically tapered joints, an external clamp is always applied to hold the two pieces together tightly and therefore provides higher supports on heavy flasks or avoids snapping and detachment of the pieces upon bending. Such ball-and-socket joining also allows the glassware to move with some flexibility to achieve the best angel to fit and join together and seal the system better. Such joints are typically labeled with two diameter numbers like ××/××, the first of which stands for the maximum diameter of the joint as the outer diameter of the ball at the base or the inner diameter of the socket at the tip while the second of which is the minimum diameter of the joint as the inner diameter of the hole in the middle of the ball or the socket. An illustrative example is shown in Figure 1. 2 for a 12/8 ball joint and a 12/8 socket joint.

Some commonly used glassware in organic laboratory is listed in Figure 1. 3.

Most glassware is used in company with clamps and stands. Support stands, also called retort stands, clamp stands or ring stands, are used to hold various labware during operations. A support stand is composed of a heave metallic rod base and a rod. Various clamps can be attached to the rod of the support stand and these clamps can then hold other pieces of equipment or glassware. Clamps are attached to the support stand using clamp holders while the adjustable ring-angle holders are one of the most used. Some clamps come in one piece with a holder for the support attachment. Different clamps are used for different labware(Figure 1. 4): Ring clamps are used to hold a ball-like glassware such as separatory funnels and regular funnels; Two or three-prong clamps are useful for flasks, such as round-bottom flask, three-neck round-bottom flasks, and Erlenmeyer flasks; Column clamps are usually specific to hold flash columns while sometimes three-prong clamps with appropriate sizes can also be used; Burette columns are designed and used for titrations; Thermometer clamps are sometimes used to hold the thermometer to monitor the temperature of a reaction; Keck clamps are used to hold two pieces of joints together. Most clamps can be adjusted to hold the labware firmly but one should avoid overtightening the clamps in order to safely manage the glassware. Some ring clamps might not be adjustable for the ring sizes and appropriate sizes of the ring clamps should be picked for certain separatory funnels.

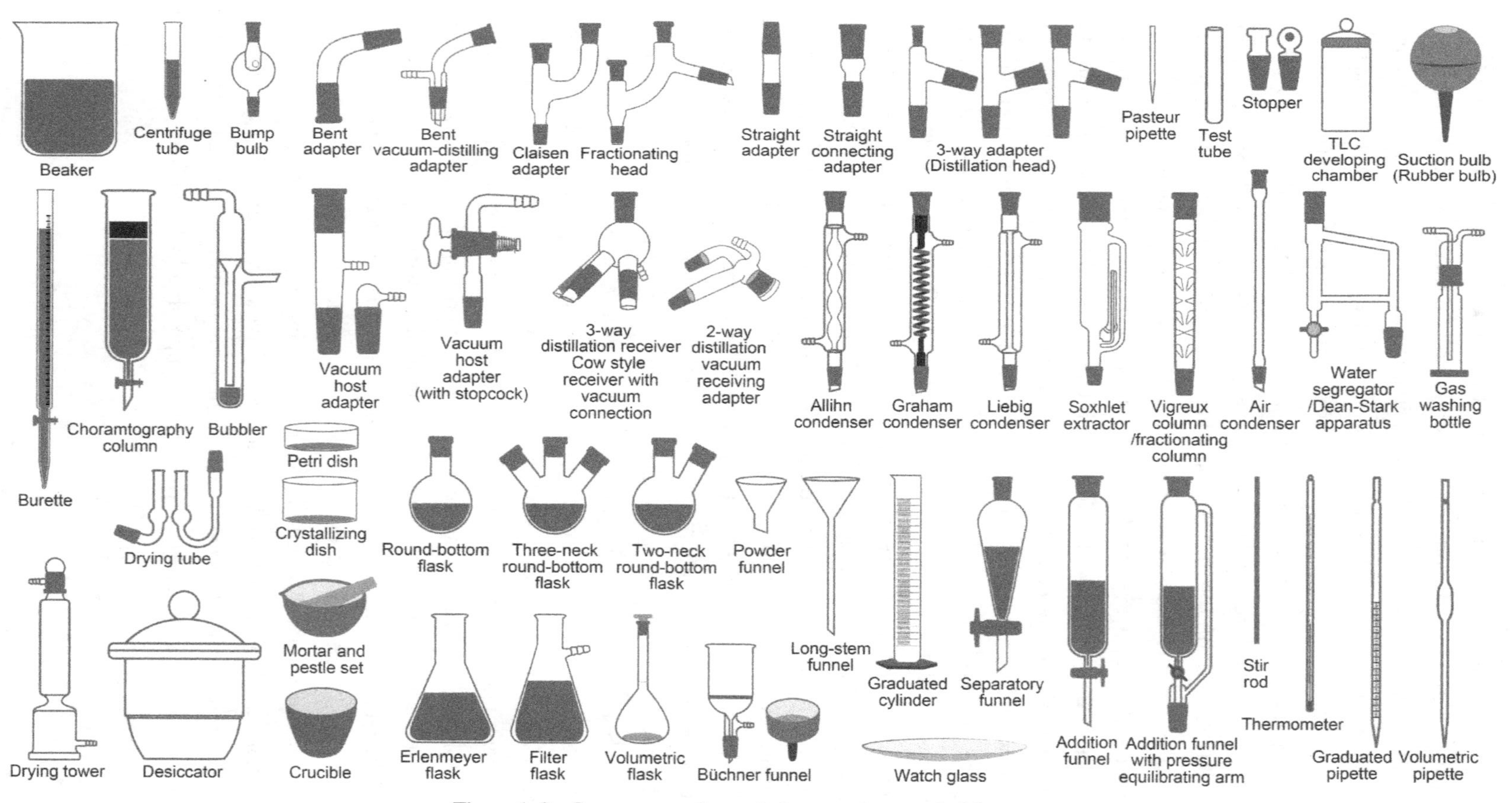

Figure 1. 3 Some commonly used glassware in organic laboratory

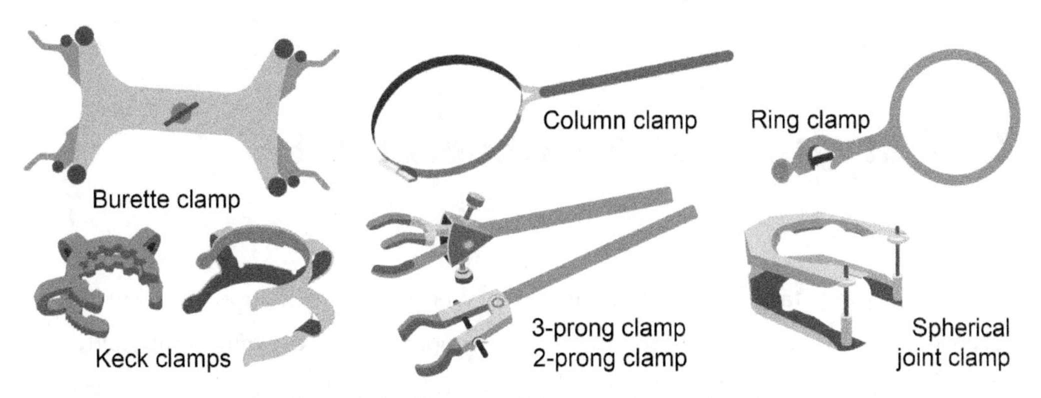

Figure 1.4 Clamps useful in organic experiments

2 Carrying out an Experiment

The very key task in an organic laboratory is to carry out an experiment for the purpose of verifying a synthetic pathway, preparing an organic compound, and investigating a reaction mechanism. There are various skills required to support an experiment. In this section, the basic techniques, such as measurement of chemicals, ways to transfer chemicals, means to achieve a certain temperature for the operation, and how to start a reaction, are briefly introduced. In addition, some important chemistry sources one might refer to are also provided in this part. ①

2.1 Measurements in a Laboratory

Measurements in an organic laboratory are not as strict as in an analytical laboratory. The measurements involved in organic laboratory typically include mass measurement and volume measurement typically. Other qualitative and quantitative measurements are not included herein.

Different measurement is applied to different substances depending on their physical states and properties. Generally, solid substances are measured by mass while liquid substances can be measured by either mass or volume. Gases can be measured by volume but in organic experiments, the gas pressure is usually measured instead.

1. Mass measurement

Electronic balances are usually used for mass measurement in the organic laboratory. Different types of scales and balances are available. To avoid contamination of the balances, weighing paper or weighing boats are used to measure solid samples, while glass containers, such as test tubes, beakers or round bottom flask, whose mass must not exceed the maximum scale of the balance, are used for liquid substances and also solid samples. Methods used on the balances are usually subtraction and taring. Both methods can be used for liquid or solid measurement. It is necessary to use a balance properly. Generally, a balance should be clean, free of dirt and on a level surface. One should use a weighing boat, weighing paper, or

---

① This part is referred to Charles D. Hodgeman, *Handbook of Chemistry and Physics*, 44th Ed. Cleveland, USA: Chemical Rubber Publishing Co. 1961, pp. 2364.

container to hold the sample in order to avoid direct contact of any sample with the balance. Even though weighing paper is used for solid substances, it is highly recommended to use glass containers or weighing boats made of inert materials to hold corrosive solid like sodium hydroxide. It is important that the container is chemically inert to the sample. When using a balance, it is necessary to avoid air movement and to have the sample or container at the same temperature as the balance for higher accuracy.

During the mass measurement, the solid sample is transferred to the weighing boat/container/paper using a spatula or a lab spoon and the liquid can be dispensed to the weighing container with a Pasteur pipette or a dropper. To avoid cross contamination, it is best not to put back any excess chemical back to the source bottle but in a waste container. Details of transferring a chemical for mass measurement can be referred to section 2. 2.

(1) *The subtraction method*

First of all, the balance should be "zeroed" by pressing the "zero" button or "tare" button for the balance to reach a "0" readout. The net mass of the weighing paper/boat or the container is then measured and the mass of the container plus the sample is measured. The mass of the sample is obtained using the equation: $m_{\text{sample}} = m_{\text{sample+container}} - m_{\text{container}}$. If the sample in the weighing container is transferred to the reaction vessel, it would be better to measure the weighing container after dispensing the sample for $m_{\text{container}}$ is actually the net mass of the container plus the residual sample therein.

(2) *The taring method*

Before the sample is transferred to the weighing paper/boat or the container, the weighing paper/boat or container is put on the balance and zeroed by pressing the "tare" button. Then the sample is directly and quickly added to the weighing paper/boat or container. If there is a long interval between the taring and measuring, or if the taring result is interrupted, a new taring process must be carried out. This is usually suitable for solids that have no remaining on the weighing container/boat/paper after transfer or the sample in the container is used directly for later operation without sample transfer.

2. Volume measurement

Volume measurement is useful for liquid samples when the density is known. Graduated measuring vessels are usually used for this purpose. Different types of measuring vessels, with different degrees of accuracy can be used depending on the amount of the liquid to be measured and types of experiments to be carried out. In the organic laboratory, graduated cylinders, transfer pipettes, and micropipettes are frequently used. Densities of most known liquid substances under standard conditions are available in *CRC Handbook of Chemistry and Physics* or in many online sources. Therefore, it is easy to get the volume to be measured for a set amount of liquid using the equation: $V=\frac{m}{\rho}$, where m is the mass of the liquid and ρ is the density of the liquid. Measure the volume of the liquid directly will be much easier and more rapid than measuring the mass. Moreover, for the solvent of a reaction, the reaction concentration is determined by the volume of the solvent used and a graduated cylinder is always used directly. It is noted that when transferring the liquid from the measuring vessel to the reaction flask, there is always a small amount of liquid remaining in the measuring vessel and the loss highly depends on the physical properties, such as viscosity. Compared to graduated cylinders, calibrated pipets and syringes are more accurate and usually applied to measure and dispense the limiting reactant while graduated cylinders are more used to measure reaction solvents or reagents that are excessive in the reaction. In addition, one needs to be cautious on the size of the graduated measuring vessels. For instance, if only 5 mL of a solvent is used for the reaction, a graduated cylinder with a scale of 10 mL is more appropriate and more accurate than that with a scale of 100 mL.

2.2 Transferring Reagents

Chemistry is a science of substance and a lot of experiments require to transfer various reagents from one container to another to some extent. Methods to transfer reagents highly depend on the physical and chemical properties of the reagent and also the requirement of the experiment. ①

1. Transferring solid substances

It is relatively easy to transfer a solid substance to an open-mouthed container, such as a beaker and an Erlenmeyer flask. The solid chemicals, when in the powder form, are easy to be transferred by tipping the solid container and slowly tapping or rotating the container to pour the solid out as displayed in Figure 2. 1. Often a spatula can be used to scoop out the solid into the target container, but the spatula must be cleaned before use, and a piece of folded weighing paper is handy for fine powder

① This part refers to Leonard B. Lane, Freeing Points of Glycerol and Its Aqueous Solutions, 1925, 17(9), 924; Roger E. Rondeau, Slush Baths, *Journal of Chemical & Engineering Data*, 1966, 11(1), 124; Alan M. Phipps, David N. Hume, General purpose low temperature dry-ice baths, *Journal of Chemical Education*, 1968, 45(10), 664; Arnold J. Gordon, Richard A. Ford, *The Chemist's Companion*, Wiley: New York, 1972; *Freezing Mixtures and Cooling Agents*, 2010, Engineering ToolBox (Available at https://www.engineeringtoolbox.com/freezing-mixtures-d_1614.html).

transferring. For some solids that form large chunks, the spatula might be used to break down the chunks into small pieces for better transfer and measurement. Similarly, a piece of weighing paper can also be used to transfer a solid from one container (the source container) to another (the reaction flask). When a large amount of solid is about to be transferred to a container with a narrow mouth, it is a good idea to use a powder funnel as illustrated in Figure 2. 2 to facilitate the transfer. If the solid is somewhat sticky, it is better to use a small amount of the reaction solvent to rinse down the solid from the weighing paper and/or funnel into the reaction flask to ensure that no loss is caused by transfer.

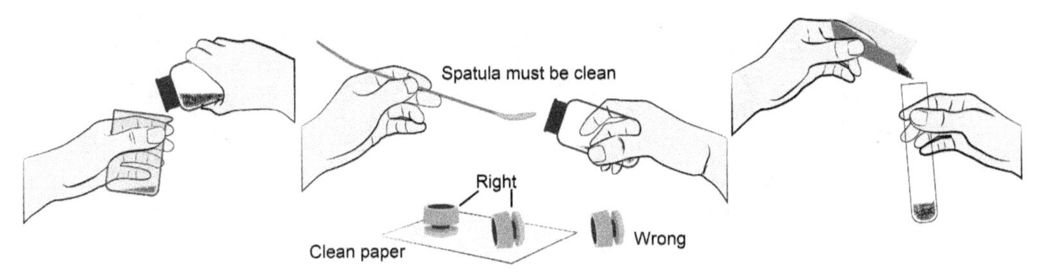

Figure 2. 1 Transferring solid reagents

One must pay attention to the physicochemical property of the solid compounds as some are vibration sensitive and it is absolutely inhibited to scratch the solid with a spatula. For this type of solid, special methods must be applied.

Figure 2. 2 Transferring chemical reagents using funnels

2. Transferring liquid substances

Liquid transfer usually relates to pouring the liquid from one container to the other. The liquid container is usually slowly tipped with its mouth touching the inner wall of the receiving container and the liquid would flow out slow. Similarly, a liquid funnel as shown in Figure 2. 2 can be used to ensure no loss of the liquid during transferring. When only a smallamount of liquid is transferred from a nearly full container, some liquid is always poured into a clean and dry beaker and a pipette is used to transfer the liquid from the beaker to the target container. One should **NEVER** put the pipette or a dropper into the source container of the liquid. However, if a dropper is provided specifically for the reagent container, one should be careful to avoid contact of the dropper tip with the receiving vessel to prevent contamination of the source bottle. In addition to transfer pipettes, Pasteur pipettes and disposable plastic transfer pipettes are frequently used in the organic laboratory for liquid transfer as shown in Figure 2. 3. For some reactions, syringes with sharp needles① are usually used to transfer liquid reagents from a sure seal bottle to the reaction vessel capped with a rubber septum and it is quite useful for enclosed reaction systems.

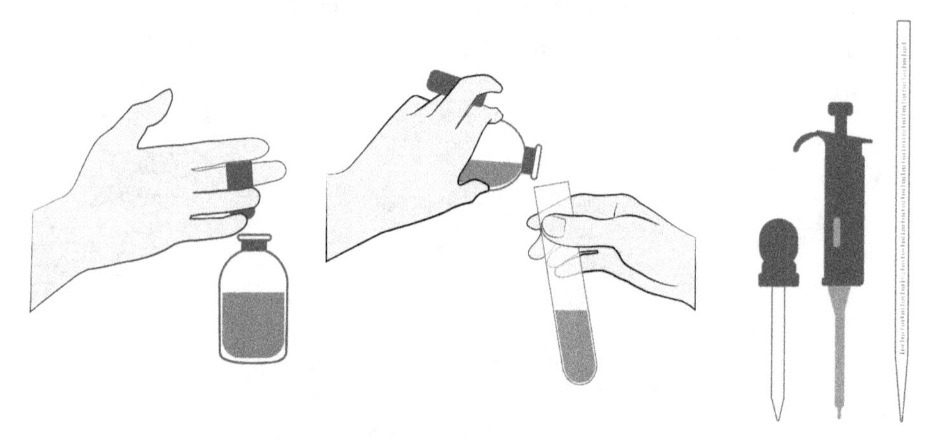

Figure 2. 3 Transferring liguid reagents

The cap of the reagent container should be placed on the bench with its inside upward on a piece of clean paper or held in the hand to avoid contamination of the cap and the container should be recapped as soon as possible when the reagent transfer is completed. It is important to hold the reagent container with the labeled side facing one's palm to avoid the corrosion or contamination of the label due to the spill of reagent and to put the container back to the shelf with the labeled side facing outward.

2.3 Heating and Cooling Methods

It is very common to undertake a reaction at a certain temperature other than room temperature. Therefore, heating or cooling operation is always used in the laboratory. Several frequently used heating and cooling methods are summarized in this section.

1. Heating

Higher temperatures can be achieved by heating. The heat source might be

---

① The disposable sharp needles should be collected in a specific sharp waste container as soon as the liquid transfer by syringe injection is completed, while the plastic syringe body should be disposed in the regular laboratory solid waste or the glass syringe should be cleaned for future use.

provided by some burners or electric heating apparatuses in a chemistry laboratory. A Bunsen burner or an alcohol burner is the commonly used laboratory equipment that is used to produce open flames by the combustion of natural gas or alcohol. Nevertheless, most organic compounds are flammable and open flames are typically restricted and disallowed in undergraduate organic laboratories for safety reasons. Instead, various heating methods involving no open flame can be used under certain circumstances.

• **Hotplates**——Hotplates using electric heating elements are widely used in chemistry laboratories and a hotplate is usually co-assembled with a magnetic stirrer for scientific research to achieve stirring and heating at the meanwhile. It can be used as a direct heat source to heat beakers, Erlenmeyer flasks, and flat-bottom crystallizing dishes, *etc*. The hotplate can also be used to apply heat to stained thin layer chromatography (TLC) plates in order to accelerate the visualization process.

• **Steam baths**——Steam baths use the hot vapor of a liquid, among which water is the most used, to heat samples in beakers, Erlenmeyer flasks, and round-bottom flasks directly to achieve purposes of warming, evaporating, or melting. A typical steam bath has a series of concentric rings, which can be assembled to fit the size and shape of the glass containers. It is regarded as one of the safest methods of heating.

• **Heating mantles**——Electric wires are usually embedded within a layer of fabric which forms a half-sphere shape that can hold a round-bottom flask. When the power of electricity, which can be adjusted by a rheostat or a variac, is applied to the equipment, a rapid increase of temperature can be achieved and containers as round-bottom flasks can be heated quickly, correspondingly. Heating mantles can usually provide a uniform heating with indirect contact of the electric heating element from the containers, which decreases the risks of glassware shattering. Accordingly, heating mantles are usually regarded as a safe and powerful heating method. In cases that the mantle is larger than the container to be heated, sand can be added to provide a better thermal contact, which function as a sand bath for the heating.

• **Heating baths**——There are several heating baths, including water bath, sand bath and oil bath as summarized in Table 2. 1. This heating bath is heated by various heating sources such as hotplates, heating mantles, or Bunsen burners. The warm materials are usually held in flat-bottom containers, such as Pyrex glass crystallizing dishes, metal pans, or heavy porcelain dishes. The container to be heated, usually an Erlenmeyer flask or a round-bottom flask, is immersed in the warm material for the samples in the container to be enveloped. The temperature of the warm material can be controlled by a temperature controller and the temperature of the samples in the flask can be monitored by a thermometer which is immersed in the materials. To achieve a specific internal temperature, the bath temperature is usually higher, roughly 10 degrees higher, than the materials in the flask.

• **Heat guns**——also called as hot air guns, can emit a stream of hot air up to a temperature range of 100 ℃ to 550 ℃ and be held by hand easily. A heat gun can provide point heating. It is most used to heat a stained TLC plate to assist its visualization.

Table 2. 1 Summary of heating baths

Bath material	Temperature range	Flash point	Advantage/Disadvantage	Risk
Silicone oil	25～230 ℃	150～350 ℃	• Provide uniform temperature • Tend to degrade over time • Slippery • Splatter when water drops into the hot oil accidently	moderate
Water	0～100 ℃	—	• No waste disposal involved • Evaporate rapidly • Dangerous when dried out	low
Sand	25～500 ℃	—	• Chemical inert and nondegradable • Easy to clean up • Provide uniform temperature	low

Among various heating baths, liquid baths might be the most used in the organic laboratory. A water bath is useful to provide a temperature lower than 100 ℃, but water sensitive materials, such as metal sodium and the Grignard reagents, must avoid a water bath. Various aqueous salt solutions can also be used to replace water for the bath to achieve higher heating temperatures. It needs to be noticed that water evaporates during the heating process and hence the water level needs to be constantly checked to avoid the dry-out of the bath as an safety risk and remained by adding water to the bath to maintain an efficient heating.

In addition to water baths, oil baths might actually be the most used in the organic laboratory and mineral oil or silicone oil is typically used for oil bath. The oil is usually regarded inert and nonvolatile. As long as they are not heated up to 200 ℃, the oil bath can be used for a long period of time. An electric heating coil is immersed in the oil bath or the oil bath is placed on a hotplate to achieve the desired heating. One should not overfill the oil bath and it is better for the oil level to reach no more than two-thirds of the height of the oil container when the container to be heated is immersed in the oil. One should also avoid the two containers (the container for the oil bath and the container to be heated) to touch each other for homogeneous and even heating. Water should be avoided to contact the hot oil bath as it may cause the hot oil to pop and splatter and become a heat hazard.

Sand baths can be used to achieve higher temperatures up to the range of 25 ℃ to 500 ℃. Sand is usually inert and nondegradable. It can also achieve even heating as the liquid bath.

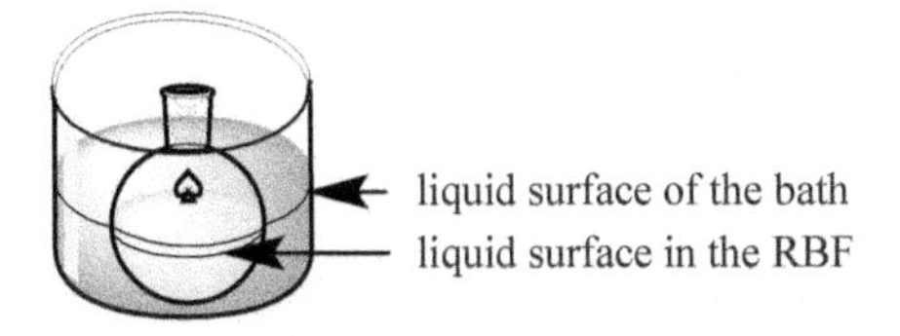

Figure 2.4 The levels of liquid surface for heating a round-bottom flask in a liquid bath

When heating baths are used, the bath liquid surface should be slightly higher than the inner liquid surface to achieve the best heating efficiency as shown in Figure 2.4. When the container is removed from the oil bath, it is better to lift the container out of the oil bath and allow the oil to drain back to the bath and cool it down to room temperature before further treatment. ①

Heating can cause skin burns and appropriate safety cautions must be raised. In addition, oil is always slippery and it may lead to accidents as slip-over.

2. Cooling

Cooling operation is usually achieved by heat transfer between the samples and the cold sources. Cooling baths are frequently used in organic chemistry. The cooling baths are always combinations of cryogenic agents and additives. The commonly used cryogenic agents are liquid nitrogen, dry ice, ice, and water. The additives can be salt or various organic solvents.

As a matter of fact, running cooling liquids is used for condensation purposes and water is one of the most used. This condensation is always achieved by using a condenser, through which the water flows. This is usually accompanied with a reflux or distillation operation.

Although the water bath is always used to cool a mixture to room temperature, the cooling efficiency varies from season to season. To get better cooling effect and lower temperatures, the ice-water bath and ice/NaCl salt bath can be used to achieve a temperature range of 0～5 ℃ or −20～−5 ℃. When other salts are combined with ice, different low temperatures can be obtained as summarized in Table 2.2.

---

① It is helpful to wipe out the residual oil on the outer surface of the container using tissue papers before further treatment.

Table 2. 2 Cooling ice baths

Additive	$m_{additive} : m_{ice}$	***T*** **(℃)**
—	Crushed ice	0. 0
H_2O	Varies	0~5. 0
$Na_2S_2O_3 \cdot 5H_2O$	1. 1 : 1	−8. 0
Acetone	1 : 1	−10. 0
Glycerol	1 : 3	−7. 0
	1 : 2	−11. 0
	1 : 1	−23. 0
	2 : 1	−46. 5
	3 : 1	−29. 5
$CaCl_2 \cdot 6H_2O$	1 : 2. 5	−10. 0
	1 : 1. 23	−21. 5
	1. 25 : 1	−40. 0
	1. 43 : 1	−55. 0
NH_4Cl	1 : 3. 33	−5. 0
	1 : 4	−15. 4
NH_4NO_3	1 : 2. 22	−16. 8
$NaNO_3$	1 : 2	−17. 8
NaBr	1 : 1. 52	−28. 0
NaCl	1 : 3	−40. 0

Dry ice and liquid nitrogen are two widely used cooling agents and can be used to provide various low temperatures in combination with additives. As the additives are usually at room temperature, the addition of additives to dry ice would result in sublimation of CO_2 or evaporation of N_2 to release a large amount of gas. Therefore, to avoid frostbite, dry ice or liquid nitrogen is slowly and carefully added to the additive in a container. If necessary, a glass stir rod is used to mix the two thoroughly to obtain a uniform cooling mixture. The formula and corresponding cooling temperatures are available in Table 2. 3.

Table 2.3 Cooling bath mixture with dry ice or liquid nitrogen

Cooling agent	Additive	T (℃)	Additive	T (℃)
Dry ice	*p*-Xylene	+13	Pyridine	−42
	1,4-Dioxane	+12	*Cyclo*hexanone	−46
	*Cyclo*hexane	+6	Acetonitrile	−46
	Benzene	+5	*m*-Xylene	−47
	Formamide	+2	Diethyl carbitol	−52
	Benzyl alcohol	−15	*n*-Octane	−56
	Ethylene glycol	−15	Di*iso*propyl ether	−60
	Tetrachloroethylene	−22	Chloroform	−61
	Carbon tetrachloride	−23	Ethanol	−72
	1,3-Dichlorobenzene	−25	Trichloroethylene	−73
	o-Xylene	−29	*Iso*propyl alcohol	−77
	m-Toluidine	−32	Acetone	−78
	Heptan-3-one	−38	Sulfur dioxide	−82
Liquid N_2	—	−196	*Cyclo*hexene	−104
	Aniline	−6	*i*-Octane	−107
	Ethylene glycol	−10	Ethyl iodide	−109
	*Cyclo*heptane	−12	Carbon disulfide	−110
	Bromobenzene	−30	Butyl bromide	−112
	Chloroform	−63	Ethanol	−116
	Butyl acetate	−77	Ethyl bromide	−119
	*Iso*amyl acetate	−79	Acetaldehyde	−124
	Ethyl acetate	−84	Methylcyclohexane	−126
	n-butanol	−89	*n*-Propanol	−127
	Hexane	−94	*n*-Pentane	−131
	Acetone	−94	Hexa-1,5-diene	−141
	Toluene	−95	*i*-Pentane	−160
	Methanol	−98		

3. A brief note on thermometers

When heating or cooling, suitable thermometers must be chosen for temperature measurement: The measured temperature must not be out of the thermometer's detection range and must not exceed the boiling point or be lower than the freezing point of the thermometer liquid. For instance, for temperatures lower than −38 ℃, the mercury thermometer cannot be used because mercury freezes. To get an accurate temperature readout, it is important to immerse all the liquid reservoir at the end of the thermometer in the measured liquid but the temperature scale should remain out of the measured liquid. In general, thermometers used in the laboratory are made of glass and fragile. They must be treated gently and carefully and the reservoir glass ball(s) should not touch the inner wall of the reaction flask or the stir bar.

4. A note on heating organic solvents

Nearly all organic solvents are volatile and flammable. It must be noted that organic solvents CANNOT BE HEATED USING OPEN FLAMES. In addition, as the electric heating wire can reach very high temperatures, heating mantles should NOT be used for low boiling and flammable ethers while heating baths would be a better choice. When such organic solvents are heated, a condenser with running water should be applied to the flask holding the solvents to avoid escape of the organic solvents into the atmosphere and the heating process should be carried out in a fumehood. To avoid overheating and drying out the solvents in the container, the power temperature of the heating source/bath must be carefully controlled.

2.4 Setting up a Reaction

To carry out an organic experiment, it is always more than simply mixing the reactants in the reaction solvent. Before the experiment is conducted, one must consider several issues: The type and mechanism of the reaction, the stoichiometric ratio of reagents, concentrations of reactants, methods to add reagents to the reaction vessel, tolerance to air and moisture by the reaction, other reaction glassware in addition to the type and size of reaction vessel, and pretreatment of the reagents and glassware. For instance, during the reaction course, it is important to keep the volatile solvents and reactants in the reaction vessel at high temperatures and this can be achieved by using a condenser. If there is some reagent that is sensitive to air, an inert atmosphere, such as argon or nitrogen gas, is usually applied to protect the reaction from air. Correct assembly of reaction glassware and apparatus and setting up a reaction are of enormous importance and discussed in this section.

1. Stirring

Generally, stirring is applied to an organic reaction mixture in order to ensure the thorough interaction among reactant molecules and reduce the reaction time. Stirring can also ensure the reaction mixture with a uniform temperature to minimize possible side reactions due to uneven temperatures or concentrations. There are several ways to stir a reaction mixture using different equipment: A stir rod, a magnetic stir bar in companion with a stir plate, and an overhead mechanical stirrer.

A glass stirring rod, stirring/stir rod, or glass rod is widely used to mix chemicals in a liquid in the laboratory and the stirring rate can be controlled manually. There are various shapes at one end of the stir rod to achieve different stirring efficiency. In addition to the stir rod, spatula, which is usually used for solid reagent transfer, can also be used to stir a mixture manually. When a reaction takes a short time and is insensitive to air or moisture, stirring by a stir rod might be used to facilitate the reaction.

Stirring by a magnetic stir bar in a reaction container on a stir plate might be the most used stirring method in organic research laboratories and undergraduate laboratories to mix mixtures of low viscosity. Most stir bars are magnets wrapped with polytetrafluoroethylene (PTFE, Teflon), which is nonflammable and stable against most reaction conditions. There are several types of stir bars in terms of shapes: cylindrical, octagon, polygon, football, egg, and fleas. Most cylindrical, octagon, and polygon stir bars are usually long and thin and are great for flat-bottom containers such as Erlenmeyer flasks and beakers, but not suitable for round-bottom flasks as they may spin irregularly. Some short cylindrical bars can fit in a round-bottom and work well. Football and egg shaped stir bars can fit well the bottoms of round-bottom flasks to achieve good stirring and are usually available for flasks of 25 mL and larger. Flea stir bars are tiny and can be used for reactions of small volumes. There are actually other shapes such as cross, crown, bone, and prism, *etc*. Each can be very useful for certain situations. Several factors must be considered in order to choose the magnetic stir bar for maximum mixing: Volume of the fluid, and shape and opening size of a vessel. Generally, a small volume of fluid require a small stir bar, which should be immersed in the liquid, while larger volumes would be matched with a larger stir bar for stronger motion and higher mixing efficiency. Shapes of vessels would help to determine the best stir bars: Flat-bottomed containers usually prefer long cylindrical stir bars while round-bottomed containers would be best to hold a curly magnetic stir bar. Though the opening of any container usually has nothing to do with choosing a stir bar, it is important to put the stir bar into the container through the opening in the first place. It is rare but possible for a vessel to have a very tiny opening and bars of small sizes are used to fit the opening. When an appropriate stir bar is selected, one needs to position the container with a stir bar in the middle of the stir plate, adjust the container to be close to the magnet of the stir plate, and slowly increase the spinning rate to achieve the best stirring.

When the magnetic stir bar is incapable of stirring a highly viscous system or a heterogeneous mixture, mechanical stirring is used. The mechanical stirring is usually achieved by an overhead stirring apparatus and it is useful for solutions of high viscosity, a large volume of solution, or heterogeneous mixtures. It is important to choose the shafts of the stir based on the size of the sample and that of the reaction vessel. During the stirring, the shafts should not touch or hit the vessel's inner wall or thermometer installed in the vessel. Furthermore, it is critical to control the spin rate by adjusting the transformer power, which is usually connected to the mechanical stirrer. As it is very powerful, it is necessary to use a specific and heavy stand to hold the overhead motor in order to minimize the potential flipping over or accidents.

2. Using clamps or flask stands to fix a reaction container

Iron stands in companion with clamps or flask stands can be used to support the reaction containers. For any reaction in a flask, an iron stand with clamps is preferred as they can provide better support to the glassware to minimize the chance of accidents. When a clamp is used, one should choose the appropriate type of clamps for the containers and try to avoid overtightening clamps in order to facilitate the later dissembling process but NOT to break the glassware. Although neck clamps are widely used to hold reaction flasks, it should not be the only support when the reaction vessel is larger than 500 mL and too heavy.

It is important to keep in mind that the orientation of the clamped flask should face the same direction as that of the base of the support stand for the best stability of the assembly to minimize the possibility for the stand to topple over as shown in the Figure 2. 5.

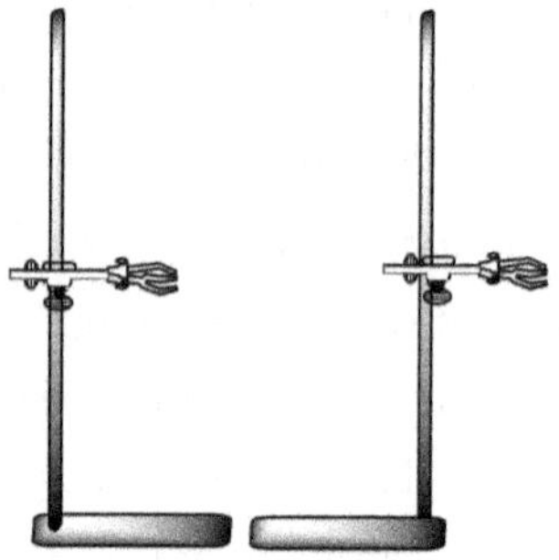

Correct Wrong

Figure 2. 5 An illustrative of clamps assembled onto the stand

3. Choosing a reaction container

All reactions take place in containers and different reactions may employ different containers. Beakers, Erlenmeyer flasks, round-bottom flasks, and three-necked round-bottom flasks are usually used in organic reactions in addition to some special reaction glassware. If a reaction uses water as the reaction solvent and the reaction takes several minutes to two or three hours, a beaker or an Erlenmeyer flask would be a good choice of a reaction container. However, when organic solvent is used for the reaction, containers with narrow opening are preferred as fitting stoppers can be used to minimize the cross talk between the reaction mixture and the open environment. When heating is involved, a condenser is always attached to the reaction container to convert the vapor back to liquid, so that the reaction container with a standard ground-glass joint must fit well directly with the condenser or using an glass adapter to achieve tight connection. If a reaction requires refluxing, measuring the reaction temperature and addition of chemicals during the reaction course, a three-necked round-bottom flask would be used for the reaction as each of the joints can fit to certain equipment, such as a condenser, a temperature adapter, an addition funnel, or a septum for syringe injection.

Besides choosing the type of a container for a reaction, the size of the container would also be important. Generally, one does not want to overfill the reaction container or use a huge container for a small volume of reaction. It is recommended to choose a container based on the rule that the solution volume is 1/3 to 2/3 of the container's volume. That is to say, for a 10 mL reaction, a 25 mL round-bottom flask would be a better choice than a 100 mL round-bottom flask. This is because a relatively small amount of reactants may stay at the inner wall of a large reaction container without thorough mixing or reacting or it has a high potential to lose the reactants or products due to production of gases or violent stirring if a small sized container is used and filled by the reaction mixture.

4. To set up a room temperature reaction without pressure accumulation

It is usually the simplest to carry out a reaction at room temperature. When a reaction with planned stoichiometric quantities is designed, one needs to set up the reaction in the laboratory. Usually, if the addition order is not an issue for a reaction, one should always try to add the solid to the container first and then the liquid reagent, and to add less active chemicals first and then more active ones, while the solvent is always added the last. After all are added to the reaction, the reaction is usually capped with a fit stopper to minimize the escape of chemicals from the reaction vessel or protect the reaction from unexpected contaminants flying into the reaction vessel if no gas is produced during the reaction course. For a reaction that produces gases, an exhaustion pathway must be applied to the reaction vessel to guide the gas either directly into the atmosphere or into a gas washing solution to consume the active volatile components before being released into the atmosphere. Then, stirring is initiated gently to achieve a thorough mixing. On the other hand, if the addition order of reagents is important, one should follow the procedure strictly to add the chemicals for the reaction. Once all has been added, one can try to monitor the reaction with various methods, such as high performance liquid chromatography (HPLC), thin layer Chromatography (TLC), and ultraviolet-visible (UV-Vis) spectroscopy, *etc.* When the reaction is complete, the active chemicals should always be carefully quenched following referred protocols, such as using alcohol to quench metal sodium before the addition of water for further workup treatment.

5. To set up a room temperature reaction under anhydrous or inert conditions

When a reaction is sensitive to moisture or air, an inert environment should be applied to the reaction. This would involve the evacuation of air in the container, filling the space with inert gases such as argon and nitrogen, and use of anhydrous solvents/reagents, *etc.* A dual manifold Schlenk line as shown in Figure 2. 6 is frequently used to alternating the evacuation and inert gas refilling process to achieve the purpose. The operation is not detailed herein because none of such operation is involved in the experiments in this book. Simply, a balloon filled of inert gas is usually used to provide an inert atmosphere and protect the reaction from the air. Besides, drying tubes and drying towers filled with drying agents are used to prevent moisture from entering the reaction vessels.

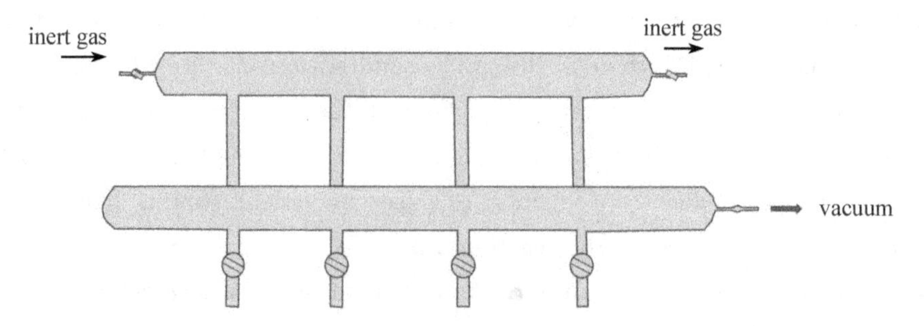

Figure 2.6 An illustrative scheme of dual manifold schlenk line

6. To set up a reaction at low temperatures

Some reactions take place at low temperatures and appropriate cooling baths should be used while not freezing the reaction mixture. The desired low temperature range can be achieved by the cooling baths summarized in Table 2.2 and Table 2.3 of section 2.3. In general, the liquid level in the reaction vessel should be lower than that of the cooling bath to realize the maximum cooling efficiency. One should also be careful on the cooling system as moisture in the air might be condensed to influence the reaction.

7. To set up a reaction at high temperatures

Similarly, various heating methods are available for reactions that require high temperatures. Heating baths can usually provide a better and more homogeneous heating to the reaction mixture as summarized in section 2.3. As most organic reactions take place in volatile organic solvents, a condenser is always attached to the reaction vessel to prevent the escape of volatiles to the atmosphere. In addition, bumping of the reaction mixture should be avoided with the assistance of boiling chips or stirring at an appropriate rate. In the meantime, the level of the reaction solution in the reaction vessel should not surpass that of the heating bath.

8. To set up an enclosed reaction with pressure accumulation

Some reactions must be undertaken at high pressures or an enclosed reaction may accumulate pressure much higher than the atmospheric pressure. For this type of reaction, a pressure resistant container or a high-pressure container must be used and specific lids and seal plugs are always paired up with the containers. Such types of reactions must be taken care of and special protection equipment, such as a pressure shield in addition to eye, skin and body protection, must be used to provide additional protections to laboratory staff as high risks of explosion or implosion are associated with the operation. For such experiments, the containers must be examined for cracks, rust, and creep before use. Distorted containers must not be used for such experiments as they are more likely to be broken during the experiments. The containers should be filled with no more than 2/3 of its volume to allow enough headspace for liquid expansion, vapor formation, and pressure accumulation. The reaction pressure must not exceed the maximum pressure limit of the container.

9. To carry out a vacuum operation

There are various operations requiring low pressures or vacuum conditions. For such operations, vacuum equipment/instruments, such as vacuum pumps, is applied to reduce the pressure of a system. It is important to have the system enclosed using fit joints and sealed connections. One thing worth addressing here is to choose pressure resistant containers as well, as the pressure difference that the container suffers from is an external force to break the container. Usually, glassware of thick wall without cracks is chosen for vacuum treatment, such as glassware used in the reduced pressure or vacuum distillation. Namely, one must check the vessels of cracks before vacuum treatment.

10. Slow addition of a reagent

Some reactions require slow addition of one or two reagents. The slow addition of a liquid reagent can be done manually with a Pasteur pipette or by an addition funnel, also called a dropping funnel. The Pasteur pipettes are usually used to add reagents to an open container in a short period of time while the reaction is generally not sensitive to moisture or air. For the addition funnel, the one with pressure equalizing sidearm is useful to handle air-sensitive reagents in an enclosed inert environment and the flow of reagent would not be influenced along the reaction course. One just needs to be slow and careful to adjust the addition rate. In addition, needles and syringes are used in pair to transfer sensitive reagents from source bottles with sure seals or to the sealed reaction vessel with a rubber septum.

Solid chemicals are usually added in small portions using a spatula with a certain time intervals. A powder funnel is recommended to be used to minimize the chemical spill.

11. Workup

It is necessary to quench the reaction and isolate the product from the reaction mixture afterwards. This process is always referred as **workup.** Even through it is not a natural part of setting up the reaction, it is somehow the most important process in an organic experiment. Most of the workup techniques are introduced in section **3** "Basic Techniques in Organic Laboratory".

2.5 Methods for Glassware Cleaning

Purity is one of the key parameters important for any chemical reagent. To minimize the uncontrolled interference from glassware used in one reaction, the glassware should be cleaned and dried thoroughly before use. One should always clean and dry glassware ahead of time instead of cleaning it right before its use while time is spent in drying the glassware in an oven.

Cleaning glassware usually requires strong detergents and running water. Scrubbing the surface of the glass with detergent and abrasive powder mixtures using a brush helps to remove deposits on the glass. Organic solvents are great to dissolve various organic tans in the containers. As the composites of dirt are complex, one should always wear gloves when cleaning glassware. At the end of the cleaning, a thorough rinse with distilled water is always recommended to prevent water spots.

The key point is to clean glassware as soon as the dirt glassware is available. It usually takes more effort to remove the same residue from glassware that has been stored for a while. Disconnect all apparatus quickly when the procedure is finished. Stopcocks and stoppers should be removed immediately from addition funnels, round-bottom flasks, and the like to avoid freezing or sticking the pieces together. Most glassware should be rinsed with compatible solvents to remove any residual organic compounds to the waste container before further washing, scrubbed with detergents or cleaning powder using a brush, thoroughly rinsed with hot water and distilled water, and dried in the oven or on a drying rack. If this cannot provide a thorough cleaning, further treatment should be applied by soaking the glassware in appropriate solutions as discussed below, which are more effective to remove the contaminants.

Sodium hydroxide alcoholic solution, also called as a base bath, is powerful to remove grease and organic contaminants from the glassware. It is important to have each piece of glassware completely filled with and immersed in the solution. Usually after half an hour or longer, the items can be taken out carefully and rinsed thoroughly with water. It might be also necessary to use brush and detergent to get rid of the contaminant from the glassware.

6 mol · L^{-1} HCl solution can be used to dissolve metal-containing compounds. When the solid dissolves, the glassware should be thoroughly rinsed with tap water, further scrubbed with detergent using a brush, and rinsed with water.

Chromic acid solution, made of dichromate and conc. sulfuric acid, is extremely powerful to degrade various organic substances and is frequently used to clean glassware especially in analytical laboratories. One should take extreme care to handle chromic acid solution because it is a strong corrosive and powerful carcinogen. Disposal of chromic acid solution must be performed properly.

There are also some other powerful washing solutions to be used in certain scenarios, such as *Aqua Regia*, acidic peroxide solution, and hydrofluoric solution. One may choose the proper one based on the type of stains and the purpose of the clean glassware.

One must keep one thing in mind during cleaning glassware: It is NOT to destroy the glassware or influence the volume of the glassware by the washing solution used. For instance, the base bath might be too basic for volumetric glassware or glass fritted funnels and it is not wise to choose the base bath to clean out these types of glassware.

After being immersed in the washing solution and scrubbed using a brush, the glassware should be thoroughly rinsed with water to remove all contaminants and detergents. At the end of cleaning, the surface of a clean piece of glassware should form uniform wetting by distilled water, provided that all contaminants are removed successfully.

When the glassware is cleaned and rinsed with water, drying process is involved to remove water to provide dry glassware for most organic reactions. At the end of any experiment, it is best to clean out all glassware used and allow it to dry in the air slowly before the next experiment starts. The glassware can be dried in an oven at no more than 140 ℃ for ~20 min and the dried and hot clean glassware should be removed with a pair of heat resistant gloves and allowed to cool to room temperature before use. The glassware can also be rinsed with some acetone or ethanol in the hood and dried in the air quickly.

Anhydrous glassware can be obtained by burning/heating the clean and dry glassware connected to a vacuum system for 5~10 min and allowed to cool to room temperature *in vacuo*. Detailed procedures are not provided herein but available in other handbooks of laboratory.

2.6 Literature Investigation for the Experiment

No experiment is coming from nowhere or because someone magically thought about it. Indeed, a thorough plan is usually involved and it is wise to refer to various sources available and relevant to each experiment in one book or the other. Even for the same type of reactions, the conditions and operation might be different and a beginner may need to study various cases to decide one's own. There are many useful books and journals to provide valuable information on experiments. One can visit the library or the online source to access the literature. Specifically, one can use reference works, chemical databases, chemistry journals, and search engines to look up information such as properties, reactivities, and reactions about chemical compounds. Some are listed below:

1. Reference works

Aldrich Handbook of Fine Chemicals, Aldrich Chemical Co., published biennially.

John R. Rumble, *CRC Handbook of Chemistry and Physics*, 100th ed., CRC

Press: Boca Raton, FL, **2019.**

Yuyu Xia, Yan Zhu, Jie Li, *Handbook of Chemical Laboratory*, Chemical Industry Press, **2015.**

Francis A. Carey, Richard J. Sundberg, *Advanced Organic Chemistry: Part B: Reaction and Synthesis*, Springer, 5th corrected ed., **2007.**

Heinz Becker, Weiner Berger, Günter Domschke, *et al.*, *Organicum-Practical Handbook of Organic Chemistry*, Elsevier Ltd., **1973.**

2. Chemical databases

Sadtler Collection of High-Resolution (NMR) Spectra, Sadtler Research Laboratories: Philadelphia.

Spectral Database for Organic Compounds, *SDBS*, National Institute of Advanced Industrial Science and Technology (AIST), Japan.

3. Chemical journals

ACS Catalysis

Advanced Synthesis and Catalysis

Angewandte Chemie International Edition

Asian Journal of Organic Chemistry

Beilstein Journal of Organic Chemistry

Bioorganic & Medicinal Chemistry

Bioorganic & Medicinal Chemistry Letters

Chem

Chemical Communications

Chemical Science

Chemistry A European Journal

Chinese Journal of Organic Chemistry

Current Organic Synthesis

European Journal of Organic Chemistry

Green Chemistry

Heterocycles

Journal of Fluorine Chemistry

Journal of Heterocyclic Chemistry

Journal of Medicinal & Organic Chemistry

Journal of Medicinal Chemistry

Journal of Natural Products

Journal of Organic Chemistry

Journal of Organometallic Chemistry

Journal of the American Chemistry Society

Natural Product Reports

Nature
Nature Catalysis
Nature Chemistry
Nature Communications
Nature Materials
Organic & Biomolecular Chemistry
Organic Chemistry Frontiers
Organic Letters
Organic Syntheses
Organometallics
Science
Science China Chemistry
SynLett
Synthesis
Synthetic Communications
Tetrahedron
Tetrahedral Letters

All journals listed herein are available online. In all of them, there might be supplemental information available and free to access for detailed experimental procedures and data.

4. Search engines

ACS Publication (www. pubs. acs. org)
ChemPort (www. chemport. org)
ChemRefer (www. chemrefer. com)
Elsevier (www. elsevier. com)
Google Scholar (scholar. google. com)
PubChem (pubchem. ncbi. nlm. nih. gov)
Reaxys (www. reaxys. com)
RSC Publishing Home (www. rsc. org)
SciFinder (scifinder. cas. org)
Web of Science (isiknowledge. com)
Wiley Online Library (online library. wiley)

3 Basic Techniques in Organic Laboratory

3.1 Assembling Glassware

To undertake an operation in chemistry, it is important for one to identify the pieces of glassware and equipment necessary for the process. After that, one needs to assemble all together to start the process. The first thing one should do is to clean up the workspace by putting away unnecessary apparatus and objects and ensure the glassware to be used is clean, dry, and without flaws. Then, one should try to assemble the glassware starting from the bottom and then upwards, and from the left to the right. The reverse order is adopted to dissemble the glassware set-up. One should clamp the apparatus onto the support stand and better position the assembly away from the edge of the lab-bench or fumehood. One may need to position the glassware with stopcocks carefully to avoid the loose detaching of the stopcock from the glassware due to gravity. All glassware should be maintained in the same plane as possible and checking leakage is always important to minimize the loss of chemicals from the glassware assembly during the experiment.

For instance, for a reaction requiring reflux shown in Figure 3. 1, one should try to position the reaction flask in the heating bath (oil bath for instance), which might be supported with a lab jack, with the external liquid level surpassing the internal liquid level. Then the flask is clamped with a 3-prong clamp (2-prong clamp can also do the job) and a clamp holder onto the support stand. Then a reflux condenser, also called as an Allihn condenser, is attached onto the flask tightly by gravity. It would be better to clamp the condenser with another 3-prong clamp to the support stand in order to secure the refluxing reaction further. The water tubing is connected to the condenser with correct inlet and outlet and the outlet tubing should be guided to the sink. When all is done, the running water should be turned on gently and both the heating and stirring can be switched on to initiate the reaction. The timing of the reaction should start with the constant dropping of liquid from the condenser back to the reaction flask. When the reaction finishes, the heating is turned off and the reaction flask and condenser are raised vertically till the flask position is above the oil bath. Until the reaction mixture is cooled back to room temperature, the running water is turned off and the condenser is taken off. The reaction flask is taken off from the clamp and its external bottom is wiped with some paper towel to remove the residual oil and the reaction mixture is then subject to further treatment.

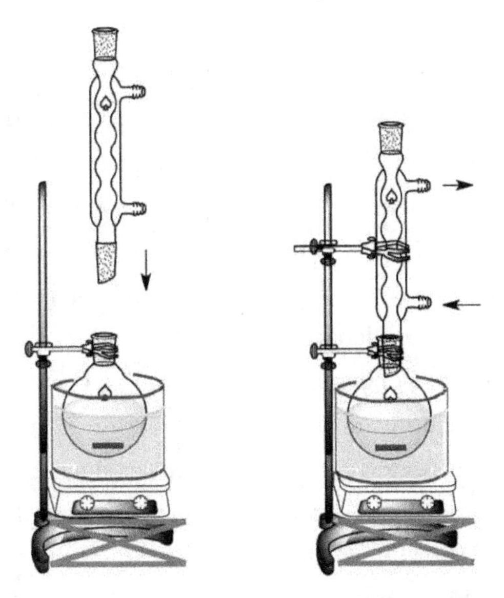

Figure 3.1 Setting up a reaction requiring refluxing

3.2 Refluxing

Many reactions and some operations, such as recrystallization, require to be heated at high temperature while the solvents and liquids in the mixture tend to evaporate. If a heated reaction system is closed, pressure would build up as vapor accumulates and this may lead to an explosion. If the reaction system is open, chemical vapor escapes to the atmosphere, which would raise critical safety issues and change the reaction concentrations and reagents' stoichiometric ratios. Accordingly, refluxing, a process to heat a mixture without losing any nongaseous reagent for a period of time from minutes to hours, is typically used. For this process, the solution is simply heated to boil and the vapor is continually condensed to liquid and drop back to the solution flask. A condenser vertically connected to the reaction flask can facilitate the heat transfer process and condense the vapor with cooling medium running in the interlayer of the condenser and guide the liquid to flow back into the flask. There are various types of condensers used for refluxing, those applying cold running solvents and others applying cool air. Usually, for reaction mixtures, the boiling point of which is lower than 150 ℃, a condenser using water flowing through its outer jacket can achieve the heat transfer for the conversion of vapor into liquid. One needs to pay attention NOT to allow too high a flow rate of the running water in order to prevent the rubber water hose from popping off the condenser to lead a minor flood. For cases when the boiling points of substances are above 150 ℃, air cooling is sufficient to condense the vapor and hence an air condenser is usually used for the refluxing.

Although condensers with appropriate cooling approaches are powerful, one should be careful on the heating rate and heating power of the heating sources. A moderate heating rate and a mild heating power is always preferred to a higher heating rate and a stronger heating power. This is because every condenser has a maximum condensation capacity to cool the vapor. If the heating rate exceeds the cooling capacity, the chemicals may as well fly away from the condenser.

Nearly all water-jacket condensers can be used to cool the vapor back to the reaction flask. The reaction condenser, or the Allihn condenser, is the most used because of its high interaction surface area for better heat transfer efficiency. In addition, when the flowing water is turned on, water must run into the water jacket from the bottom to the top of the condenser: Water flows into the bottom inlet and comes out at the top outlet of the condenser. This can ensure the water jacket is filled with water and all air bubbles would be kicked out.

When the refluxing is carried out using a reaction condenser, it is good to control the heating rate/power and the flow rate of the running water to prevent the vapor from rising up to the 1/3 height of the condenser and better to keep the vapor-liquid conversion level at the first half-ball of the reaction condenser.

Aside from tap water used as the cooling matrix, refrigerated circulators accompanied with various cooling mediums are nowadays used in research organic laboratories for condensing purposes with reliable temperature controls. For these refrigerated circulators, water and ethylene glycol are usually used as the cooling matrix and temperatures down to −15 ℃ can be achieved. This can provide stronger heat transfer effect and is quite useful for highly volatile solvents, such as diethyl ether.

Some reactions are moisture sensitive and refluxing must be done under anhydrous conditions. As the condenser attached to the reaction flask has an open mouth to the atmosphere, one can connect a drying tube containing drying agent to the condenser. As for air sensitive reactions, one can either flush the reaction system with inert gas or cap the condenser with a septum, which is pierced through using a needle and attached with an inert gas filled balloon, to achieve an atmospheric reaction under inert conditions.

3.3 Simple Inert Gas Protection

Inert gases, such as argon and nitrogen, are usually used to protect a system that is air sensitive or moisture sensitive. The system can be filled with inert gas using a dual manifold Schlenk line as shown in Figure 2. 6 in section 2. 4 of "Setting up a reaction" by alternating between vacuum and inert gas. It is possible that the lab lacks a Schlenk line and one can flush the container with inert gas thoroughly to remove the residual gas completely. To flush inert gas through the container, usually a gas inlet and an outlet should be applied, as shown in Figure 3. 2. In addition, a balloon filled with inert gas can also be applied to the reaction flask to provide an inert atmosphere. If one is dealing with highly air sensitive compounds or reactions, one can carry out all operations in a lab glovebox, which is specifically designed for delicate operations.

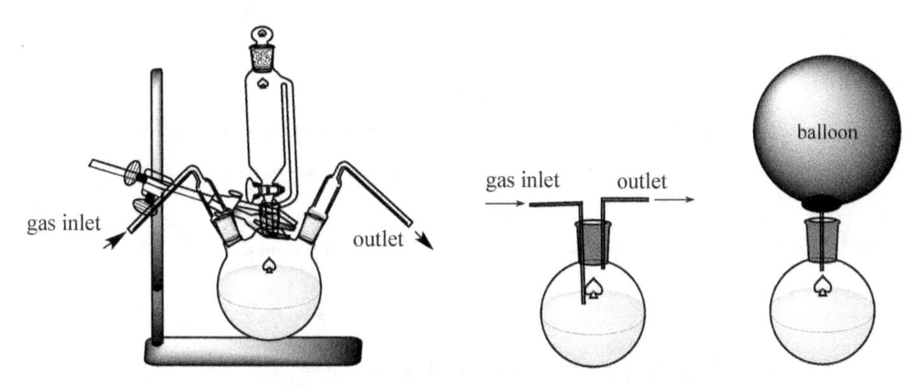

Figure 3. 2 Simple ways for inert gas protection

3.4 Smelling a Chemical

Odors, are one of the important physical properties a substance holds. When a new product is produced, it would be important to get an idea on how it smells. Although nowadays the odor of a compound is no longer so important, notorious substances are less likely to be used in cosmetics. On the other hand, just like color change, odor change is also accompanied along many reactions. Nevertheless, one should never try to take a deep breath directly from the container in order to smell the content because this would raise high risks of destroying the mucous membranes, air ways or lungs. Instead, one should try to minimize the amount of chemicals breathed in. To do so, one should try to breathe in just enough chemical to get the sense of the smell. A standard method to achieve this is called the wafting technique. One should first cup one hand above the container with the cap off, fan the air toward one's nose, and breathe shallowly for the smell as shown in Figure 3. 3.

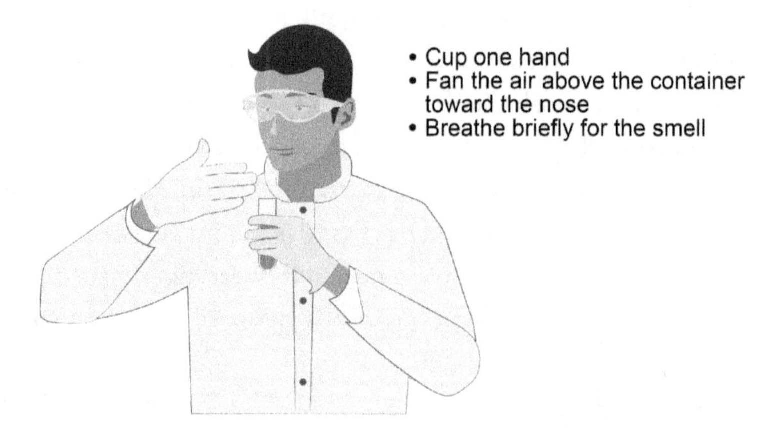

Figure 3.3 An illustrative for the wafting technique to smell minimum chemicals

Note: *As the bioactivities of many organic compounds are unknown, one should be cautious on smelling these chemicals.*

DO NOT put your nose directly onto any container with compounds.

3.5 Boiling Point and Distillation

All matters have three phases: Gas, liquid, and solid. The phases are mainly determined by the pressure and temperature the matter is subject to. Curves on the diagram as shown in Figure 3.4 demonstrate the transition between two phases, such as between liquid and solid states, between gaseous and liquid states, and between gaseous and solid states. The three curves usually come across at a triple point, which stands for the coexistence of all three states of the matter. When enough molecules spread out and escape the liquid phase to form gaseous vapor, bubbles can form and boiling takes place. Two factors to determine when the bubble forms are the external pressure and temperature. The temperature stands for the energy the molecules must absorb to convert to the gaseous state from the liquid state, while only the accumulated vapor pressure equals to the external pressure that a bubble breaks to allow the vapor escaping from the bulky liquid. The temperature, at which the liquid boils, is referred as the boiling point. Therefore, generally, at a place where the atmospheric pressure is lower, such as at

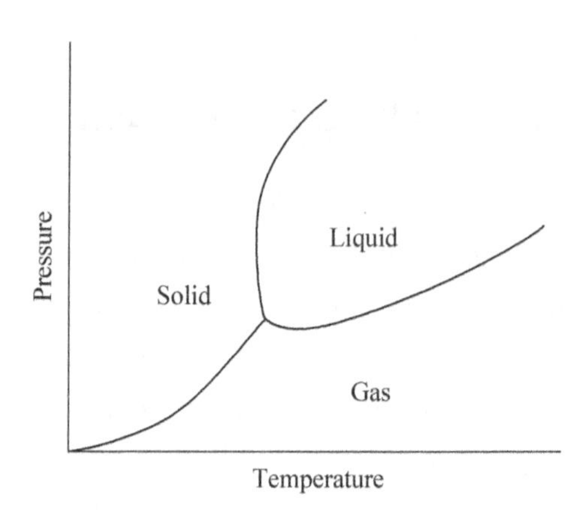

Figure 3.4 A typical phase diagram

the top of a mountain, the boiling point is usually lower than that on an ocean.

The boiling point of a compound is determined by the intermolecular interactions. Typically, polar compounds usually have higher boiling points than nonpolar compounds as there are stronger intermolecular interactions among polar molecules. Boiling points of some commonly used organic solvents are summarized in Table 3. 1.

Table 3. 1 Boiling points of some organic solvents

solvent	Hexane	MeOH	EtOH	Et_2O	THF	DCM	EtOAc	AcOH	DMF	DMSO	PhH	PhMe
bp (℃)	69	64. 7	78. 4	25. 0	65. 4	39. 8	77. 4	118. 2	153	189	80. 1	110. 6

As the boiling point is the highest temperature a liquid can be heated to and it is the temperature of the vapor when the vapor pressure equals the external pressure, measurement of boiling points usually relies on measuring the temperature of the liquid or the vapor when the phase conversion reaches an equilibrium at boiling. Either immersion of the bottom of a thermometer to the liquid or incubation of the bottom of the thermometer in the space that vapor would occupy is used. The former usually uses a refluxing set-up while the latter is formally applied in a distillation set-up.

In addition to measuring the boiling point of a liquid compound, distillation is widely used to separate and purify liquid mixtures. As a matter of fact, distillation is one of the important laboratory techniques making use of the phase transition when a liquid boils under a certain atmosphere. Usually, either atmospheric pressure or reduced pressure is applied to the distillation system to achieve the separation of liquid mixtures. One should be aware that a liquid substance usually boils at lower temperatures under reduced pressures. In a typical distillation, several pieces of glassware are used: Distillation flask, distillation head, thermometer with an adapter, a Liebig condenser, a receiving adapter and a receiving flask. Depending on the pressure source that the receiving adapter is connected to, there are **simple distillation** (also called as atmospheric distillation) and **vacuum distillation.** The former is connected to the atmosphere and the latter is connected to a vacuum system such as an oil pump for a certain reduced pressure. Especially when vacuum distillation is undertaken, one must use vacuum resistant glassware, which has no cracks. In addition to simple distillation and vacuum distillation, fractional distillation is more powerful for liquid mixtures with difference in boiling points of components less than 20 ℃ under an atmospheric pressure and an additional fractional column is used to achieve better separation of components.

When setting up a distillation, the distillation flask should contain a liquid with a volume no less than one third and no more than two thirds of the flask volume. Then either boiling chips or a stir bar with appropriate shape and size is placed in the flask.

All glassware should be assembled from the bottom to the top and from the left to the right. One should use clamps to provide necessary support for the assembly and check joint connection to avoid leakage before heating is applied. When vacuum distillation is used, vacuum grease may be applied to the joints for better vacuum and protection of the system from air flying in.

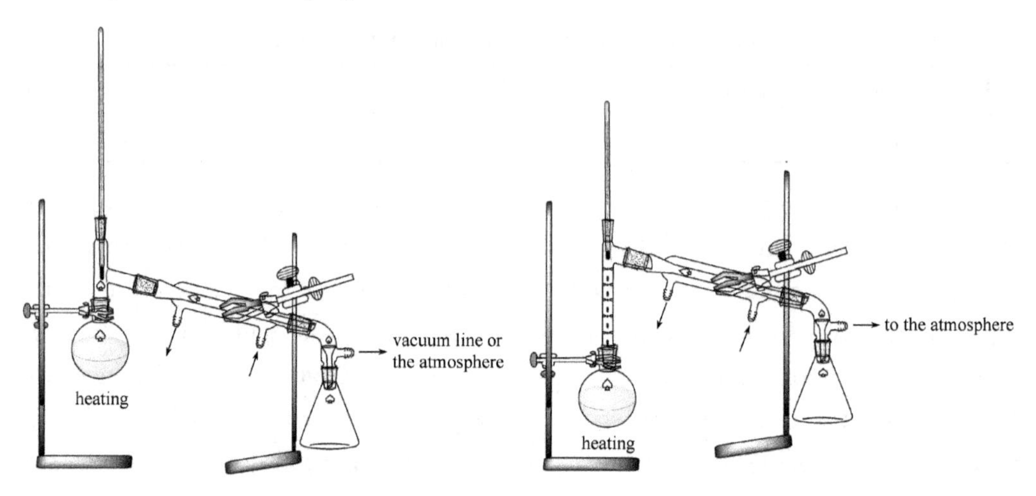

Figure 3. 5 Distillation set-ups. The simple or vacuum distillation (Left) and the fractional distillation (right)

If one plans to collect several components, two at one distillation for instance, 3-way distillation receiving adapters can be used to collect the fractions from the one with the lowest boiling point to the one with the higher boiling point subsequently and into separate collection flasks without disconnecting and changing receiving flasks along the distillation course.

For all these distillation types, it is necessary to record the boiling points for the fractions that are collected while the actually pressure applied in a vacuum distillation should also be recorded. It is difficult to know the exact boiling point for every chemical at different pressures. At this case, one can refer to a nomography to estimate the boiling point of a compound at a certain pressure if its boiling point at the standard atmospheric pressure is known. There are three scales for a nomograph: The boiling point temperature scale, the boiling point temperature at sea level scale and the general pressure scale. In order to use the nomograph, one can connect two known values using a ruler and read the third scale along the line the ruler demonstrates.

Fractional distillation, on the other hand, is a technique used to separate a mixture of two or more liquid components that cannot be isolated by simple distillation. The fractional distillation uses a fractionating column to achieve multiple re-distillation. In the fractionating column, packing or spinning band is usually used to maximize the contact between the ascending vapor and the descending liquid while such contact always allows a better fractionation. With each vaporization-condensation equilibrium cycle, the vapors are enriched with a certain component and only the most volatile vapors stays predominantly in the gas form and travels to the top of the column. Better separation can be achieved with fractionating columns and a number of such phase equilibriums are involved, which usually are referred to more theoretical plates. The Vigreux distillation column is one of the most used fractionating columns in organic laboratories.

Steam distillation is another type of distillation that is used for thermosensitive compounds. It is different from the distillation mentioned above and does not follow the Raoult's rule. In general, the two components involved in the steam distillation are usually immiscible to each other at their liquid states. The organic compounds separated by this technique is labile at high temperatures and prone to decompose and should have some vapor pressure no smaller than 1.3 kPa at 100 ℃. What happens is the steam flows from the distillation flask to the collection flask and it carries a small amount of vaporized compounds to the collection flask. Given enough time, all the vaporizable components can be carried to the collection flask. Typically for a steam distillation, water is boiled to produce the steam and either it is boiled *in situ* or the steam is guided to the distillation flask that is used for the steam distillation. To achieve this, one end of the glass tube should be immersed in the solution and be positioned as close as possible (~8—10 mm) to the bottom center of the long-neck round bottom flask. It is also important to tilt this flask at an angel of ~45° against the horizontal line to minimize the potential solution spilling to contaminate the distillate. The solution to be distilled should be less than 1/3 of volume of the long-neck flask. The set-up of steam distillation is demonstrated in Figure 3.6. Not only the steam-production flask should be heated, but also the long-neck flask holding the solution to be distilled should be heated to a lower temperature to avoid the accumulation of condensed water in this flask. On the other hand, the flow rate of the condensing water should be adjusted carefully to prevent the accumulation of solidified product and blockage of the condenser.

At the very beginning of the steam production, the T-shaped glass tube should be open to atmosphere. When the water in the flask is boiled and a large amount of steam is produced, the valve of the T-shaped glass tube should be closed and the steam distillation starts. When there is no more oily distillate collected, the distillation is complete and the valve of the T-shaped glass tube should be switched to open to atmosphere before the heating for steam production is terminated.

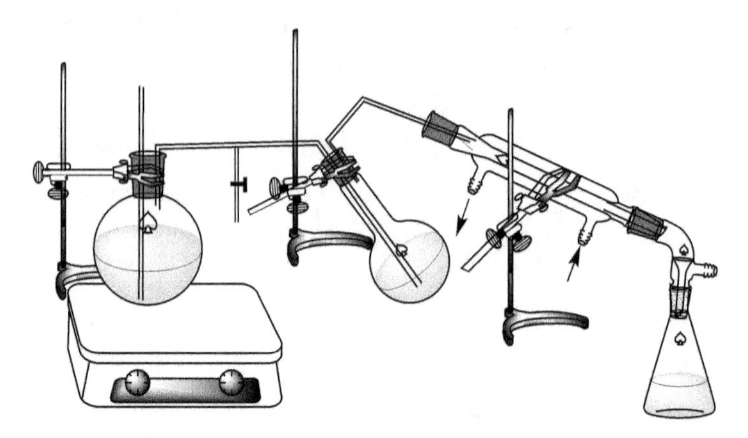

Figure 3.6 A set-up of steam distillation

A simpler set-up of steam distillation that is convenient for an undergraduate laboratory is demonstrated in Figure 3.7. A separatory funnel or an addition funnel (non-pressure equilibrated one) is used as the water reservoir and water is slowly added to refill the vaporization flask. The vaporization flask is assembled with a regular simple distillation set-up and the vapor is then condensed by the condenser and collected in the collection flask.

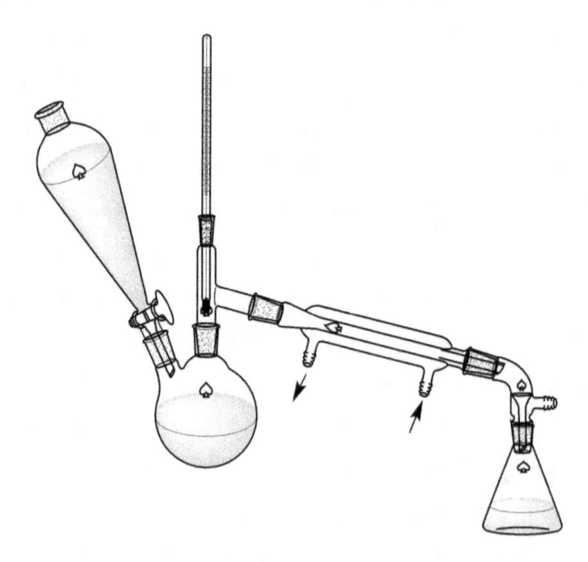

Figure 3.7 A simple set-up for steam distillation in which the steam is produced *in situ*

3.6 Melting Points

Melting point is a characteristic physical parameter for a compound and it is the temperature at which the phase of substance changes from a solid state to a liquid state. For a pure compound, it always provides a clear, sharply defined melting transition. During the melting process, all the energy absorbed by the substance is consumed as heat of fusion, and the temperature remains constant when the solid phase and liquid phase coexist. The melting point of a pure compound is only dependent on the pressure it is recorded. Trace amount of impurities would lead the melting point to change and also enlarge the melting ranges dramatically. The measurement of melting point is relatively simple, reliable, rapid, and cost-effective. There are many methods to measure the melting point. Among them, the capillary method is usually used. In this method, a thin glass capillary tube with one end sealed is filled with a compact column (~2—3 mm high) of the compound to be tested. The capillary tube is subject to a heating source in close proximity to a high-accuracy thermometer. The melting process of the solid should be closely observed and the temperature at which the solid starts to turn into liquid and that at which the solid completely converts to liquid should be recorded as the melting point range. To achieve a good melting point determination, the sample should be dry and in a fine powder form.

The heating source can be provided by an oil bath in a Thiele tube (or called b-tube) or by a heated metal block which equipped in a regular melting point apparatus. Usually, the temperature in the heating source should be slowly increased until the sample in the tube transits into the liquid state. The second try to measure the melting point of a sample should be waited until the heating source cools to at least 20 ℃ below the expected melting point.

If one is trying to measure a compound with a known melting point to check its identity and purity, one can heat the heating source with a fast rate and slow down the heating rate until 20 ℃ below the expected melting point. This can ensure the melting equilibrium is reached and one would not miss the melting point.

If one is working with a new compound, the melting point of which is not yet reported anywhere, one may heat the sample with a reasonable rate to obtain a rough but inaccurate melting point readout, and then follow the protocol to measure the melting point more accurately as if the melting point of the compound is known as described above.

To fill the sample in a glass capillary melting point tube

1. A dry and powdered sample should be used for a melting point test. If the solid is in large chunk or granular, one should pulverize/smash the solid into fine powder and possibly dry it further before the melting point measurement.

2. Use a clean glass capillary melting point tube with one end sealed.

3. Try to use the open end of the capillary tube to capture the solid sample by jabbing it into a pile of the solid, then quickly inverting the capillary tube, tapping the

tube gently on the bench top, and drop the tube down several times with the open tube facing up through a long narrow tube to pack the solid at the closed end of the capillary tube.

4. If necessary, step 3 is repeated to load the sample until a sample height of 2～3 mm in the tube is packed.

Noted: Too much sample in the tube would result in an artificially broader melting range. Inappropriate packing, leading the sample in the tube to be too loose, can cause the sample to collapse and shrink when heating, which could interrupt the judgement for the melting temperature.

3.7 Sublimation

It is well known that the phase transition of a pure substance always goes from solid to liquid or from liquid to gas, or *vice versa*. In addition, as shown in the phase diagram demonstrated in section **3.5**, it is possible for the phase transition between the gas phase and the solid phase, such as that a solid compound evaporates directly to form the vapor or the gas form condenses into a solid without going through the liquid phase. This usually requires the substance to have an appreciable vapor pressure and temperature below its triple point. For instance, iodine, dark purple crystals, can easily turn into purple vapor upon stand, and dry ice can turn into colorless carbon dioxide gas quickly at atmospheric pressure. The process that vapor is obtained without melting the solid compound is called sublimation. This is used as a separation of a sublimable substance from insublimable impurities and a deposition process is involved to condense the vapor into the solid phase to collect the pure matter. Only substances with the following natures can be purified by sublimation: 1. It is solid and vaporizes without melting; 2. It should be thermostable and not decompose; 3. Its vapor can be condensed back to the solid phase upon cooling; 4. The impurities do not sublime. Sublimation may take place at atmospheric pressure but more appreciably at reduced pressure.

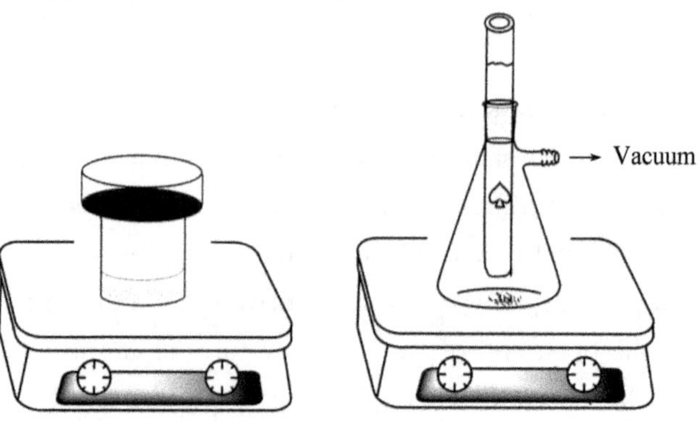

Figure 3.8 An illustrative of simple sublimation apparatus

A sublimation is usually carried out using a sublimation apparatus, which has a vessel for the evaporation of the sample and a cooled surface referred as a cold finger to condense the vapor to pure solids. Two illustrative set-ups are shown in Figure 3.8 as the left is for sublimation at atmospheric pressure and the right for vacuum sublimation.

3.8 Filtration

Filtration is one of the important techniques used to separate solids from liquids. It can be used to collect the solid product from a reaction mixture or recrystallization mixture, and remove any insoluble solid impurities from a solution, including the solid drying agent such as anhydrous sodium sulfate and magnesium sulfate intentionally added earlier in order to remove the residual water in the solution. Gravity filtration and vacuum filtration are commonly used in organic laboratory. The former uses a combination of conical funnels and Erlenmeyer flasks while the latter uses a combination of Büchner or Hirsch funnels and filter flasks. In general, filter paper with appropriate sizes or shapes is required for both filtrations. During filtration, the solid is retained on the filter paper and the liquid going through it and being collected is called the filtrate. Similarly, a Pasteur pipette that is packed with silica gel or alumina can also be used to remove impurities at a certain point.

The filter paper with a medium relative filtering speed and of particle retention sizes larger than 8 μm works well and is usually used for both gravity filtration and vacuum filtration. In addition to the filter paper, cottons or glass wools can also be used to fill the conical of the funnel to replace the filter paper for filtration while finely powdered inert solids such as Celite can be used in companion with filter paper to remove smaller powdered solid impurities in the solution.

1. Gravity filtration

Filter paper can be folded for conical funnels in gravity filtration. There are generally two types of folding to facilitate the filtration as shown in Figure 3.9. Either can be used to collect the solid or the filtrate.

(1) *Folding a filter cone*

First fold the round filter paper in half and crease it. Then fold again and crease the filter paper into a quarter circle. Separate one outer layer of the paper from the other three and squeeze slightly the outer surface of the paper at the creases to form a conical shape, which is later placed into a conical funnel.

(2) *Folding the fluted filter paper*

Fluted filter paper is more popular nowadays as it provides more surface for solid-

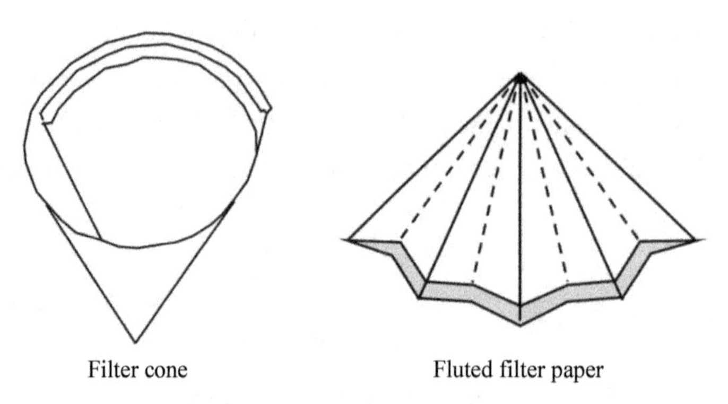

Figure 3. 9 Folding of filter paper

liquid separations and accelerates the gravity filtration. This is especially useful for hot recrystallization to remove insoluble impurities while the crystallizing is prevented during filtration. To fold a fluted filter, a regular round filter paper is creased in half four times. Then each of the eight newly presented sections in the filter paper is folded inward and last the paper is opened to form a fluted cone. The fluted filter paper is also commercially available.

2. Vacuum filtration

Vacuum filtration can use a conical funnel. Typically, a conical funnel is first filled with glass wool or cotton and fit onto a filter flask with a rubber adapter before vacuum is applied for the filtration.

But funnels with flat plates having pores or tiny holes are more frequently used in vacuum filtration and are usually made of porcelain. Büchner funnels (Figure 3. 10) or Hirsch funnels are of this kind. These funnels are characterized by the sizes in diameters. For instance, a Büchner with ϕ4. 3 cm (the diameter of the funnel is 4. 3 cm) can be used to collect the solid of 0. 2~1. 0 g via filtration. To successfully carry out a vacuum filtration, filter paper with appropriate sizes must be used for the funnel and it should lie flat on the plate and ought to cover all the holes or pores in the plate without curling up the side. A rubber adapter or a wood stopper with a hole to fit the funnel neck is used to attach the funnel with the filter flask.

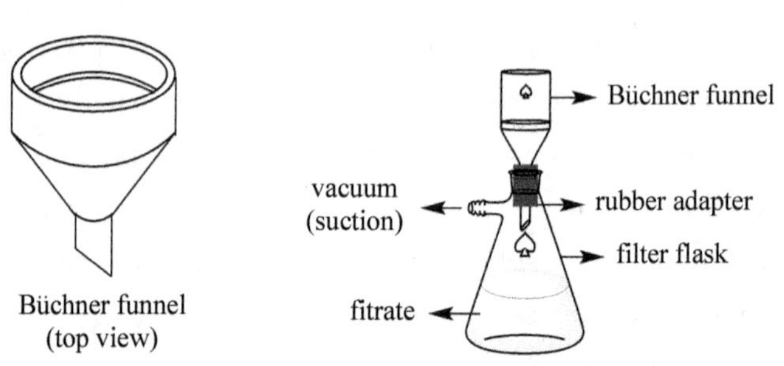

Figure 3. 10 Büchner funnel and set-up of vacuum filtration

During vacuum filtration, one should be careful not to overfill the filter flask or liquid surpassing the level of the vacuum branch will flow into the vacuum source or in the trap flask.

To better perform a vacuum filtration using a Büchner funnel, a piece of pre-cut filter paper with an appropriate size is placed on the funnel and a small amount of solvent present in the mixture is used to wet the filter paper. Then the vacuum is connected to the filter flask to pull the paper tightly onto the funnel. The mixture to be filtered is immediately poured into the funnel. At the end of the filtration, the vacuum is broken by disconnecting the filter flask with the vacuum line before the vacuum source is turned off to avoid the liquid in the pump to flow back into the vacuum flask. To prevent the accident of tilting over the filtration apparatus, one should hold the flask firmly with one hand and use the other to disconnect the vacuum line to break the seal, or fix the filtration apparatus onto an iron stand.

In addition to filtration, it is also worthwhile mentioning the other liquid-solid separation techniques briefly. One is decantation and the other is centrifugation.

3. Decantation

Decantation is a process to pour carefully and gently the liquid away above the particles present in the mixture. Usually, the large, heavy solid particles settle down in the bottom of the container and the top liquid is easy to be poured off to another container. This technique is not suitable for a liquid mixture containing a large number of solid or fine particles. To apply this technique, the solid materials must be settled down and the supernatant is carefully poured to another container, during which a stir rod is usually used to help the flow of the liquid (Figure 3.11).

Figure 3.11 Decantation

4. Centrifugation

Centrifugation is an effective method for liquid-solid separation. It is powerful for liquid samples containing suspended particles, for which filtration usually results in poor separations. Centrifugation is also very helpful for emulsion systems to facilitate the layer separation. Centrifugation requires a centrifuge and all sample tubes must be balanced before the operation.

3.9 Recrystallization

Not only liquid always presents in mixture, but solid compound is also sometimes contaminated with impurities. To ensure the usage of the compound and avoid ambiguous effect of impurities, compound of high purity is always required. Especially from some organic reaction, the target molecule is usually accompanied with by-products and un-reacted precursors. The compound is useful only after it is isolated in a pure form. Hence, various techniques to purify compounds have been developed. Distillation mentioned earlier is one of the frequently used methods for isolation of pure liquid compounds. The one frequently used to purify solid compounds from a crude solid is **recrystallization** and in many events it is called **simple crystallization.** This is to dissolve a crystalline material or a powder mixture or even a paste or slurry in a minimum amount of hot solvent and under the cooling process crystals with mainly one component crash and precipitate. This is an extensive and useful method to obtain pure solid compounds in organic laboratories and also in chemical industry.

There are two types of impurities: Those more soluble in a given solvent than the main component and those less soluble. By taking full advantage of such solubility difference, the solid substance can be isolated and purified. In an ideal case, one solvent can completely dissolve the compound to be purified at high temperature, usually the boiling point of the solvent, and the compound would be completely insoluble in that solvent at room temperature or even lower. In addition, the impurity would be either completely insoluble in that particular solvent at high temperatures or extremely soluble at low temperatures. In the former case, the impurity can be removed by filtration at a high temperature, and for the latter, the impurity remains completely in solution upon cooling. In either case, a small amount of compound being purified should remain in solution at low temperatures and form a saturated solution. In general, low solubility at low temperatures minimizes the loss of purified compound during recrystallization.

The solvent is critical for recrystallization. A good recrystallization solvent is expected to satisfy that the solubility of the crystals in the hot solvent is much higher than that in the cold solvent. In addition, a suitable recrystallization solvent should usually be partially volatile in order to be easily removed from the purified crystals. The solvent should be inert to the compound being purified and it'd be better to boil below the melting point of the compound to be purified to avoid solid to melt before dissolving (oiling out). The boiling point of the solvent should also not be too low, otherwise either the solvent evaporates quickly or the crystallization occurs too rapidly. In selecting a good recrystallization solvent, one should also consider flammability, toxicity, and expense.

The rule of "like likes like" can be followed for solvent selection in recrystallization. A polar solvent dissolves polar compounds and a nonpolar solvent dissolves nonpolar compounds. The most commonly used recrystallization solvents are listed in Table 3. 2.

Table 3. 2 Some solvents commonly used in recrystallization

Solvent	Formula	Polarity	Boiling point* (℃)
toluene	$C_6H_5CH_3$	slightly polar	110. 6
water	H_2O	very polar	100

(**Continued**)

Solvent	Formula	Polarity	Boiling point* (℃)
acetonitrile	CH_3CN	polar	82.0
*cyclo*hexane	C_6H_{12}	nonpolar	80.7
benzene	C_6H_6	nonpolar	80.1
ethanol	CH_3CH_2OH	polar	78.4
ethyl acetate	$CH_3CO_2C_2H_5$	slightly polar	76.7
hexane	$CH_3(CH_2)_4CH_3$	nonpolar	68.7
tetrahydrofuran	$(CH_2)_4O$	polar	66
methanol	CH_3OH	polar	65
acetone	CH_3COCH_3	polar	56.5
dichloromethane	CH_2Cl_2	slightly polar	40
diethyl ether	$(CH_3CH_2)_2O$	slightly polar	35

* The boiling point is at one atmosphere

If a single solvent (though preferred in recrystallization) cannot achieve good purification, a solvent pair is often picked. The solvents must be miscible, like water-ethanol, acetic acid-water, and ether-acetone, *etc.* Typically, the compound being recrystallized is preferred to be more soluble in one solvent (the good solvent) than the other (the poor solvent). The compound is dissolved in a minimum amount of the hot good solvent and a certain amount of the poor solvent is slowly added until a turbid mixture is resulted while boiling. The mixture can then be cooled down for crystallization.

In general, only oversaturated solution can result in crystallization of the solute. While crystal formation requires a nucleation core, it is always helpful to add several crude crystals or small crystals of the compound as seeds to the saturated solution at high temperature to accelerate the crystal formation. Or sometimes scratching the wall of the container using a stir rod may also result in some crystal from the glass to drop in the solution as seeds for recrystallization. The crystal would grow on the core and solute molecules would try to fit into the crystal lattice. If the growing process is slow enough, the molecules can dissolve back in and deposit to the crystal lattice to fix any defects for better and more perfect crystals, which also means purer and larger crystals. On the opposite, if the crystal growing process is rapid, there are always traps of impurities within the crystal lattices and smaller and finer crystals are resulted in. As a resulting, during a recrystallization operation, after the saturate solution at high temperatures (usually the boiling point of the solvent) is obtained, the solution is allowed to cool to room temperature or lower by itself without any assistance in order to obtain better shapes and higher purity of the crystals. A quick drop of temperature by immersing the hot solution in cold water bath is not recommended.

When one plans recrystallization, one should estimate the amount of solvent to be used in order to determine the size of the container used for recrystallization and probably try to choose a container two or three times more than the volume of solvent one plans to use. Sometimes, one may have a rough idea on the amount of solvent to be used on the basis of the solubilities of the solute in the chosen solvent at different temperatures. One should focus on the solubility at the boiling point of the solvent in order to determine the approximate amount of solvent that might be used as the saturated solution at high temperature always means an oversaturated solution at lower temperature and this can enforce the crystallization to take place. In such a case, one can add 1/3 or 1/2 of the calculated amount of solvent to the condenser-attached flask (round bottom flask or Erlenmeyer flask can be used) containing the solid to be recrystallized and heat the mixture to boil and reflux. Then the leftover solvent is added slowly in small portions to the flask while boiling until all solid disappears. The mixture is then allowed to cool to ambient temperature for crystallization. One may find that a bit less or more solvent might be used than the calculated amount. To avoid solution bumping, boiling chips or stirring should be introduced to the recrystallization mixture.

If one has no clue about the solubility of a compound in a solvent, one should be carefully and slowly add the solvent using a Pasteur pipette into the compound while the mixture is boiled to reflux and also ensure to use as minimum solvent as possible to dissolve the solid all.

If the impurity is insoluble in the solvent, one should try to follow the general recrystallization step to dissolve the solid by adding solvent in small portions until when 1 or 2 more milliliters of solvent is added but results in no further dissolution of the solid. At this point, the mixture should be quickly filtered to remove the insoluble impurity and allow the filtrate to cool naturally and crystallize.

When the recrystallization mixture cools to room temperature and the crystal formation completes, one should try to collect the crystals by filtration and rinsing the recrystallization flask with minimum of the recrystallization solvent to transfer the residue solid into the funnel. The crystals collected by filtration is usually wet and must be dried thoroughly further. Drying in vacuum is always recommended for organic compounds as the residual solvent can be removed quickly while no risk of fire is associated if an organic solvent is involved in comparison with being dried in an oven.

Brief summary of a recrystallizing procedure

(1) Heat and dissolve the solid with minimum amount of an appropriate solvent (or solvent pair)

(2) Remove insoluble impurities as soon as the desired compound dissolves

(3) Allow crystals to grow by cooling the solution slowly to an ambient temperature

(4) Collect crystals by vacuum filtration (recommended) or gravity filtration

(5) Rinse the crystals with a small amount of cold solvent

(6) Dry the crystals with appropriate methods

3.10 Drying Organic Solvents

There are various reactions requiring an anhydrous environment and therefore it is necessary to use dry reagents at the first place. For instance, when organolithium reagents are used, such as BuLi, anhydrous solvents must be used because BuLi reacts quickly with water. As a result, the solvents are usually pretreated by special drying procedure while various drying agents as shown in Table 3.3 are used to form crystalline hydrate with water molecules present in the reagent.

Table 3.3 Some commonly used drying agents

Drying agent	Properties	Capacity	Comment
H_2SO_4	Acidic	High	Aggressive and oxidative
P_2O_5	Acidic	High	Aggressive and usually not in contact with solvent
Molecular sieves	Slightly acidic	High	Reusable, suitable for different solvents based on the pore sizes
K_2CO_3	Basic	Moderate	Not suitable for acidic compounds
KOH	Basic	Very high	Good for amines
$CaCl_2$	Neutral	Medium to high	Not suitable for alcohols
$CaSO_4$	Neutral	Low	General drying agent
$MgSO_4$	Neutral	High	Good general drying agent
Na_2SO_4	Neutral	Very high	Good general drying agent

The inorganic salts listed in Table 3.3 are usually characterized by their capacity, efficiency, and speed to dry the solvents and one must be careful to determine which drying agent is used for a certain solvent. Indeed, all these salts are actually water soluble and hence they are only useful when a trace amount of water is present in the solvent and the drying agent is inert to the solvent. For instance, it is not wise to use calcium chloride to dry ethanol as calcium chloride preferentially forms complexes with ethanol instead of forming hydrated crystals. For another example, even though magnesium sulfate is regarded as a neutral salt, it is slightly acidic in presence of water. As a result, magnesium sulfate is actually not good to dry acid sensitive reagents.

The amount of the drying agent to be used is determined by the water content in the solvent and more water present usually requires higher amounts of the drying agent. One should add as much as it needs to dry the solvent and the drying agent quickly forms hydrated crystals which clump together at the bottom of the container. Too much drying agent would do no harm to the solvent but may lead to loss of the solvent at the solid-liquid separation step. So a general guide for the amount of drying agent is used: Small portions of a drying agent are added[①] to the solvent in an Erlenmeyer flask and the flask is swirled to mix the drying agent with the solvent until there is some drying agent freely moving around while swirling the flask. When the liquid is dry, it should appear clear and a small amount of the drying agent in the flask remains the particle size and appearance of its original form. The drying process usually takes around 20 minutes and the liquid-solid mixture is then separated by gravity filtration or decantation.

Molecular sieves are porous materials and can be used to trap either water or solvent molecules. They are widely used to dry gases and solvents. For instance, 3 Å molecular sieves are useful to dry ethanol and unsaturated hydrocarbons. Recently, it was reported that by simply passing the solvent through a column of molecular sieves or silica gel, anhydrous solvents with water contents of ppm levels can be achieved.

Such drying process usually can provide a solvent containing a very little amount of water. If anhydrous solvents are required, more strict treatment is required by using reactive metal, metal hydride, or P_2O_5 to consume the residual amount of water in the solvent and a distillation system is usually assembled later to collect the anhydrous solvent, which is then either directly used or stored in the presence of molecular sieves. This is common in various research laboratories and not in the scope of most undergraduate laboratory experiments.

3.11 Drying Organic Solutions

Some organic solutions, such as separated from an extraction treatment, may contain some dissolved water. It is important to remove the residual water for further operations. The large amount of water present in the solution can usually be removed by treating the solution with an equal volume of saturated sodium chloride aqueous solution, also referred as *brine*, if the solvent is immiscible to water. The salt in the water is capable of pulling the water from the organic phase to the aqueous phase. Then by using a separatory funnel, the two layers can be easily and carefully separated. After this, the routinely used drying agent, such as anhydrous sodium sulfate, magnesium sulfate, calcium chloride, calcium sulfate, or potassium

① Usually a spatula tip of the drying agent is added one time.

carbonate, *etc*. is added to further dry the organic layer as described in section **3.10.**

During the filtration or decantation step to separate the solution from the solid drying agent, one should use additional fresh solvent to wash the drying agent to minimize the loss of the product.

3.12 Drying Solid Organic Compounds

A lot of organic compounds are in the solid state and they can be isolated from various methods, such as precipitation, recrystallization, chromatographic separation, and sublimation, *etc*. To use it in a variety of characterization and additional conversions, it is important to get rid of the residual solvent accompanying with the solid. To do this, a drying process is usually involved and removal of the solvent by evaporation is always the key. To achieve this, one can choose several ways:

High vacuum is frequently used to dry organic solids① with high efficiency. The solid sample can be placed in a vacuum resistant container, such as a round bottom flask, and the container is then connected directly to the vacuum line for an hour or longer. The solid sample can also be held in a beaker, which is placed in a vacuum desiccator that is connected to a vacuum line for the same purpose. This is suitable for not only solid organic compounds but also for liquid organic compounds as long as the compounds are not volatile or sublimable.

Drying in the air is slow but also useful. Usually, one can spread the solid sample on a watch glass and place it in the fumehood to allow the evaporation of the residual solvent. This may take hours to days to dry the sample depending on the nature of the solvent. One should be careful to protect the sample in the watch glass from contamination or overthrowing.

Flowing air or gases can accelerate the drying process by taking away the solvent vapor and it is used in certain case to dry solid compounds. In addition, increasing the environment temperature can largely enhance the rate of drying, and so an infrared drying lamp, microwave irradiator or oven is commonly used to dry thermostable compounds.

Note: One should decide which drying method is used based on the nature of the compound and the availability of instrument in the laboratory.

3.13 Extraction

Extraction is a widely used method separating a substance out from a mixture. It involves the removal of a component of a mixture by contact with a second phase. Extraction

① The solid should not be sublimable.

is based on the compound partitioning in two immiscible phases and obeys the "like likes like" principle. Solid-liquid and liquid-liquid extractions are commonly performed by batches and continuous processes. The removal of caffeine from coffee beans with dichloromethane is an example of a liquid-solid extraction. Crystal violet may be removed from a water solution by liquid-liquid extraction with *n*-amyl alcohol (pentan-1-ol).

1. Liquid-liquid extraction

Liquid-liquid extractions are somehow often used as a key purification procedure in organic laboratories. Some common applications of liquid-liquid extractions involve the following:

(1) Collection of organic compounds from the aqueous solution using an appropriate organic solvent

(2) Isolation of products from an organic reaction

(3) Removal of acid, base, and salt impurities

(4) Removal of organic acids and bases from other organic compounds

Liquid-liquid extractions indeed involve partitioning of a solute, ***A***, between two immiscible solvents, ***S*** and ***S'***. This distribution between the two layers at equilibrium may be described by the following relationship.

$$K_p = \frac{concentration\ of\ \boldsymbol{A}\ in\ \boldsymbol{S}}{concentration\ of\ \boldsymbol{A}\ in\ \boldsymbol{S'}}$$

The partition coefficient K_p is usually larger than 1. So, ***S*** is a better solvent for ***A*** to dissolve. The K_p may be used to evaluate the effectiveness of an extracting solvent and to plan an extraction.

A suitable extracting solvent can facilitate the extraction and save a lot of effort in separation. To choose an extracting solvent, one should obey the general selection rule:

(1) Immiscible to the original solvent

(2) Having a favorable K_p

(3) Nonreactive (with the exception of aqueous solutions of acids and bases)

(4) Different in density from the original solvent

(5) Relatively volatile and easy to be removed from the solute

(6) Less toxic

Usually, an extraction takes place by using a solvent pair of water and an organic solvent. The commonly used organic solvents for extraction are listed in Table 3.4.

Table 3.4 Some commonly used organic solvents for extraction

Solvent	Et_2O	CH_2Cl_2	$CHCl_3$	$CH_3(CH_2)_4CH_3$	EtOAc	$PhCH_3$
Boiling point (℃)	35	40	61.2	69	76.7	110.6
Density ($g \cdot mL^{-1}$)	0.71	1.32	1.42	0.66	0.9	0.87

One of the most commonly used pieces of equipment for extraction is a separatory funnel, or "sep funnel". Products of many organic reactions are expected to be collected in the organic phase from the aqueous layer, which can remove salts, acids or bases. Even in some events, the reactions require that one component be removed by acid-base extraction. In addition, the isolation of useful compounds from naturally occurring materials relies on the liquid-liquid extraction. All of these can be accomplished in a 'sep funnel'.

As shown in Figure 3. 12, a sep funnel is a pear-shaped glass device equipped with a stopcock at the bottom and a stopper to close the top opening. Sep funnels come in all sizes from about 50 mL to 5 L or larger. *They are all very FRAGILE and should be handled carefully.*

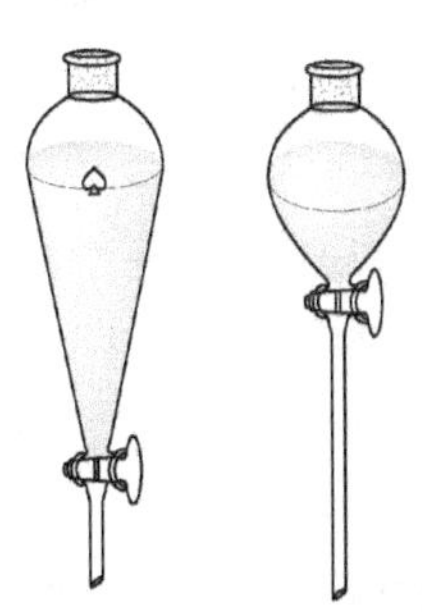

Figure 3. 12 Sep funnels

The sep funnel acts like a cocktail shaker. With the stopcock closed, ingredients are added through the top. The stopper is securely placed, and the contents are shaken and mixed thoroughly. While the sep funnel is being shaken, the stopper must be held securely in place and the stopcock must be tightly shut.

Quite often, the components evaporate to develop some pressure upon shaking. This pressure must be **carefully** relieved by slowly opening the stopcock while the sep funnel is inverted. *The sep funnel must always be 'aimed' away from any nearby person to avoid incident when the pressure can be larger than expected.* After the sep funnel has been shaken a few times and the pressure is relieved several times, the sep funnel is placed upright in an iron ring, and *the stopper is removed.* ① When the two layers are clearly separated, the lower layer should be carefully drained into an Erlenmeyer flask by slowly opening the stopcock, and the top layer is then poured out via the top of the sep funnel.

If a second extraction is needed, the layer to be extracted is placed in the sep funnel and a new portion of the extracting liquid is added. Then the shaking, inverting, and venting process is repeated. This is commonly encountered during the "workup" of a chemical reaction. In addition to the extraction with fresh organic solvent, the organic layer can also be washed with several different kinds of aqueous solutions.

Reminder: The stopcock must be cleaned before and after each use. When not in use the stopcock is left clean and assembled, but *very loose fitting. The plastic will distort if left tightly held in the glass part.* When in use, the stopcock nut must be tightened to keep the stopcock securely held in the glass, and a leakage test must be performed for the sep funnel before use.

2. Liquid-Solid extraction

Another useful extraction technique is liquid-solid extraction. It uses a Soxhlet extractor (Figure 3. 13) as the key equipment. In generally, the solid sample to be

---

① one may wait to remove the stopper right before the operation to separate the layers. This can allow a minimum loss of then organic solvent from evaporation.

extracted is placed in the Soxhlet extractor and a condenser is attached to the Soxhlet extractor after it is attached to the solvent reservoir round bottom flask, which is heated to vaporize the solvent. The solvent vapor is quickly condensed back to the Soxhlet extractor once it reaches the condenser and the solid is then immersed in the solvent. When the liquid surface in the extraction chamber surpasses the top point of the siphon arm, all solution in the chamber quickly flows back to the boiling flask due to the siphon phenomenon. The solid sample in the Soxhlet extractor chamber is extracted into the solvent and only a limited amount of solvent is used to achieve a maximum extraction for its continuous re-use in the system. This technique is useful for compounds that have limited solubility in the solvent and such compounds will be enriched in the boiling flask. To use this technique, there is usually some components in the sample that are not soluble in the solvent.

Figure 3.13 A Soxhlet extractor

3.14 Rota-evaporation

Rota-evaporation is short for rotary evaporation, which involves a high surface area of a thin solvent film along the rotation of a flask. This type of evaporation is highly useful to remove volatiles and concentrate a solution. To accelerate the evaporation of solvent, this operation requires a rotary evaporator connected to a vacuum system, such as a water aspirator or a water pump. In general, a round bottom flask of thick wall is attached to the rotary evaporator and is spinning while the vacuum is turned on. The spinning round bottom flask would result in a thin film, which can help the liquid film heated evenly. The vacuum applied by the vacuum system allows the evaporation of volatiles at lower temperatures and this ensures a rapid evaporation of the volatiles as well. To avoid the bumping of solutions into the rotary evaporator, one can use a bump bulb with appropriate standard ground-glass joints to fit in the round bottom flask. To carry out a successful rota-evaporation, the round bottom flask should usually be filled with no more than two thirds of the flask volume while the external heating bath, usually water bath, is heated to a temperature lower than half of the boiling point of the volatile. It would be helpful to control the spinning speed as well. A faster spinning speed is always associated with a quicker heat exchange for the film and faster evaporation is general achieved, but it is much easier to cause liquid bumping in the flask. Therefore, one should be careful on the spinning rate. Moreover, the vapor should be condensed by the condensation system of the rotary evaporator. Running tap water is usually used as the cooling matrix while a cooling circulator filled with either water or other cooling liquid is much more effective and popular to minimize the solvent to enter the environment nowadays. A cartoon illustrative scheme of rota-evaporation is shown in Figure 3. 14.

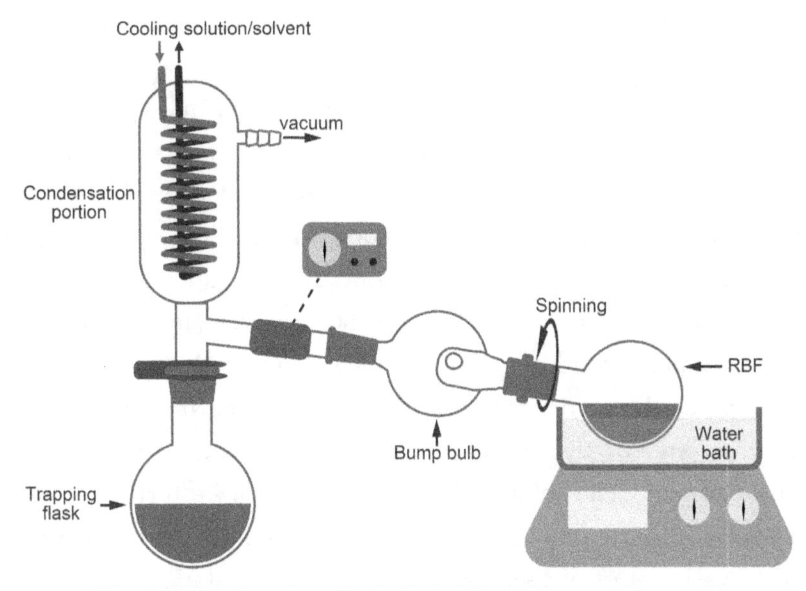

Figure 3. 14 A cartoon illustrative of the rota-evaporation set-up

Reminder: One should first check if there is enough free volume for the receiving flask to collect the volatile and turn on both the vacuum system and the cooling system before rota-evaporation. It is necessary to ensure there are no flaws or cracks on the flask used for rota-evaporation as reduced pressure is usually associated. Erlenmeyer flasks should not be used for rota-evaporation as they are usually not pressure resistant and can be a potential hazard upon uneven pressure treatment between the interior and exterior. Both the flask and the bump bulb should be cleaned before use. It is highly recommended that keck clicks are used to hold the joints tightly together to avoid accidently falling over of the rota-evaporating flask. The spinning should be turned off and the vacuum should then be destroyed by opening the valve to the atmosphere before the vacuum pump is terminated. One should either recycle the collected solvents or dump them into the corresponding waste solvent cans, halogenated or non-halogenated, as soon as the rota-evaporation ends.

3.15 Chromatography

It is important to identify if a substance is pure or not, because impurity may raise serious issues in the follow-up reactions and side effects in biological systems.

Based on this, a separation plan can be proposed or revised. In chemistry, there are various ways to achieve qualitative analysis of purity. One of the methods is chromatography, which is an analytical method based on separation. Chromatography depends on partitioning chemical components between two phases, usually one stationary phase and one mobile phase. The mobile phase usually travels through the stationary phase, which remains where it is. Along the movement of the mobile phase, different chemical components demonstrate distinct mobilities because they have different interaction strength with the mobile phase and the stationary phase. This difference in mobilities leads to the separation of the components. As the relative mobilities for different components in the same system only depend on the interactions between the compounds and the mobile/stationary phase, the separation can also be used to qualitatively analyze the compounds for purities and identities.

Based on different ways to carry or support the stationary phase, chromatography can be divided into planar chromatography and column chromatography. The planar chromatography has the stationary phase loaded on a planar support, such as thin layer chromatography, or the planar support itself is the stationary phase, such as paper chromatography. The column chromatography can be further categorized into gas chromatography and liquid chromatography according to the type of the mobile phase. Both have varieties of kinds with different types stationary phases. A quick summary is demonstrated in Table 3. 5.

Table 3. 5 Categories of chromatography

<table>
<tr><th colspan="3">Chromatography</th><th>Mobile</th><th>Stationary</th></tr>
<tr><td rowspan="2">Planar Chromatography</td><td colspan="2">Thin layer chromatography (TLC)</td><td>Liquid</td><td>Silica gel or aluminium oxide on glass or plastic plates</td></tr>
<tr><td colspan="2">Paper chromatography</td><td>Liquid</td><td>Paper</td></tr>
<tr><td rowspan="8">Column Chromatography</td><td rowspan="2">Gas chromatography (GC)</td><td>Gas-solid chromatography</td><td>Gas</td><td>Solid</td></tr>
<tr><td>Gas-liquid chromatography</td><td>Gas</td><td>Liquid</td></tr>
<tr><td rowspan="6">Liquid chromatography (LC)</td><td>Flash chromatography</td><td>Liquid</td><td>Silica gel or aluminium oxide</td></tr>
<tr><td>Adsorption chromatography</td><td>Liquid</td><td>Solid on underivatized support</td></tr>
<tr><td>Ion-exchange chromatography</td><td>Liquid</td><td>Support derivatized with fixed charged residues</td></tr>
<tr><td>Exclusion chromatography</td><td>Liquid</td><td>Porous and inert support</td></tr>
<tr><td>Affinity chromatography</td><td>Liquid</td><td>Support with specific immobilized ligand</td></tr>
</table>

3.15.1 Thin-Layer Chromatography

Thin layer chromatography (TLC) is one of the neat methods that are frequently used in an organic laboratory to characterize organic compounds for purity and identities but requires no high-tech instruments. It can be used to isolate a small amount of organic compounds as well. TLC has also been used in monitoring reaction progresses. It can provide key information that would guide the flash chromatography for compound separation in a larger scale (grams or even kilograms). TLC only requires very tiny amount of sample and reveals the result quickly in only several minutes. Two elements are involved in TLC analysis: The stationary phase and the mobile phase. The most used stationary phases are silica gel and aluminium oxide. The stationary phase is usually coated on a glass surface or aluminium surface. In most cases, an inert inorganic green fluorescent indicator, called green fluorescing F_{254}, is co-mixed in the stationary phase to provide better visualization of compounds on the plate under UV lights. On the other hand, the mobile phase can be any solvent or solvent combination. This is usually referred as the solvent system. The most used solvent systems are hexane/ethyl acetate and dichloromethane/methanol. But petroleum ether, diethyl ether, tetrahydrofuran, chloroform, toluene, acetone, ethanol, and water are also used for TLC analysis while ammonium water, acetic acid, or triethylamine may be added to the developing solvent in certain cases. This is based on the combination of a relatively nonpolar solvent and a relatively polar solvent. With different percentage of each pair, we can obtain solvent systems with different polarities, which are suitable for different compounds with various polarities to move along the stationary phase. The principle of TLC analysis is based on intermolecular interactions between the compounds of interest with the stationary phase and/or the mobile phase, and after TLC development, compounds with different polarities will show different mobilities on the TLC plate (stationary phase). If the compound is pure, only a single spot would be present on the TLC plate after the development in several suitable solvent systems.

To undertake a TLC analysis, the compound/mixture must be loaded on the TLC plate using a capillary tube first. Therefore, the compound/mixture is first dissolved in a suitable solvent (volatile solvent preferred) and spotted on the TLC plate gently. Then the TLC plate is dried and placed in a chamber (TLC chamber) containing the solvent system and developed as shown in Figure 3.15. What needs to mention is that the sample loaded on the TLC plate must be higher than the liquid surface of the developing solvent to prevent the compound from dissolving into the solvent. The solvent would "climb" along the plate due to capillary effect until the solvent front reaches the level that is lower than the top edge of the plate and this process is called development of the TLC plate. In order to minimize the effect of solvent evaporation on the plate, the air in the TLC developing chamber is usually saturated with the solvent vapors and this can be quickly achieved by inserting a developing-solvent wetted filter-paper in the chamber prior to the TLC chromatogram development.

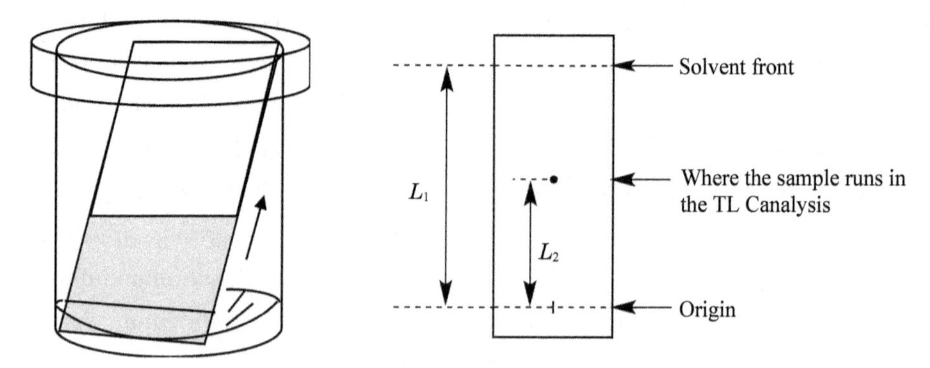

Figure 3. 15 An illustrative of developing a TLC plate and marks for the origin and the solvent front of the TLC chromatogram

Both the places where the sample is spotted before the development (origin) and where the solvent front reaches when the development is terminated must be marked with a pencil and the distance between the solvent front and the origin (L_1) and that between the sample spot after development and the origin (L_2) are measured in order to calculate the retention factor (R_f) using the equation $R_f=\frac{L_2}{L_1}$. The R_f value for a certain compound is independent of time or chambers only if the stationary phase (type and thickness) and the solvent system are consistent as well as the temperature and external atmospheric pressure. Overloading the sample on the plate may result in tailing spots and unrepeatable R_f's.

R_f values would be a good guidance to determine the developing solvent. Usually, a solvent system that can afford R_f values in the range of 0. 3～0. 6 is regarded as an appropriate developing solvent. However, there is no universal solvent system in TLC analysis. One has to test if a solvent system is suitable for the sample. A general principle would be to use a polar solvent for polar compounds while to use a nonpolar solvent for nonpolar compounds. This implies that polar compounds may not move when using a nonpolar developing solvent system but nonpolar compounds would move as fast as the solvent and reach the solvent front by a polar developing solvent system. Based on this, for a sample of unknown polarity, one can possibly start with a low polar developing solvent system and if the R_f value is too small, a more polar solvent system should then be tested. This should be repeated until an appropriate developing solvent system is achieved. Such trial-and-error process is actually quite common in chemistry. One more thing worth mentioning is that silica gel plates are not happy with highly polar solvent and methanol or ethanol is not used with a concentration higher than 20v/v% in combination with other nonpolar solvent. If 20% methanol or ethanol in dichloromethane does not move the compound, for instance, some additives, such as triethyl amine or acetic acid, should be added to adjust the acid-base properties of the developing solvent.

It must be noticed that the sample must be visualized by a certain method. If the compound contains chromophores and hence is regarded UV active, the UV light at 254 nm is usually used to visualize the compound on the TLC plate, where a dark spot will be viewed with a green fluorescent background under the UV irradiation, which is due to the fluorescent indicator F_{254} loaded on the support of the commercially purchased silica TLC plates. Compounds with poor or no UV activities must be visualized by other methods, such as iodine stain and potassium permanganate stain, as listed in Table 3. 6. Most stains use a dipping technique and the soaked TLC plate may require further heating by a heat gun to allow the visualization. For instance, stain of amines using ninhydrin stain solution usually requires heating to accelerate the amino group to react with ninhydrin to give the colored product for visualization.

Table 3. 6 Some commonly used TLC stains

TLC stains	Components	Phenomena upon staining
Iodine	Iodine in silica gel	Temporary stain with unsaturated compounds
Ceric sulfate	$Ce(SO_4)_2$ in sulfuric acid solution	Give various colors for different compounds
2,4-Dinitrophenylhydrazine	2,4-Dinitrophenylhydrazine in sulfuric acid aqueous-alcoholic solution	Orange spots for aldehydes and ketones
Ferric chloride	$FeCl_3$ in ethanol	Blue spots for phenols/enols
Ninhydrin	Ninhydrin in butanol and HOAc	Blue to yellow sports for primary and secondary amines
Permanganate	$KMnO_4$ in K_2CO_3 aqueous solution	Yellow spots for unsaturated or alcoholic compounds

A summary of TLC procedure

1. Get a precoated TLC plate and gently mark the origin that should be ~0. 5 cm from the bottom edge of the plate using a pencil.

2. Spot the plate at the line with a small amount of dilute solution to be analyzed using a capillary tube.

3. A filter-paper is folded in a way to fit in the developing chamber and the suitable solvent is added. After the chamber is capped, allow the filter-paper to be wetted and the chamber atmosphere is saturated with the solvent vapors.

4. Place the sample loaded and dry TLC plate in the chamber with the head marked with pencil headed down. Do not touch the filter-paper with the TLC plate or

have the solvent surface surpasses the marked line on the TLC plate.

5. Develop the chromatogram until the solvent front reaches 0.5~1.0 cm away from the top of the plate.

6. Remove the plate from the chamber and mark the solvent front immediately using a pencil.

7. Visualize the spots on the chromatogram and outline each spot using a pencil.

8. Measure the distances and calculate the R_f value for each spot.

As TLC analysis is highly useful to monitor a reaction and analyze a mixture, it would be really helpful to load reference compounds, such as the starting materials and products if available, onto the same TLC plate with the sample at different places on the marked origin line.

3.15.2 Flash Chromatography

One of the most used separation techniques in organic synthesis is flash chromatography. It is actually a rapid form of preparative liquid column chromatography, which requires no delicate pump system, injector, degassing apparatus, or fancy detection system as regular liquid chromatography. It uses medium to short cylindrical column that can be filled on site quickly with the stationary phase, such as silica gel or aluminium oxide, and applies air pressure to push the liquid mobile phase to pass through. Because silica gel or aluminium oxide is also used as the stationary phase in thin layer chromatography, flash chromatography has high consistency with thin layer chromatography and thin layer chromatogram is always used to determine the liquid mobile phase for flash chromatography used to purify the sample. Flash chromatography is powerful, rapid, and efficient to separate compounds in the scale lower to milligrams and up to tens of hundreds of grams. Components with a R_f difference lower to 0.15 can be separated using flash chromatography.

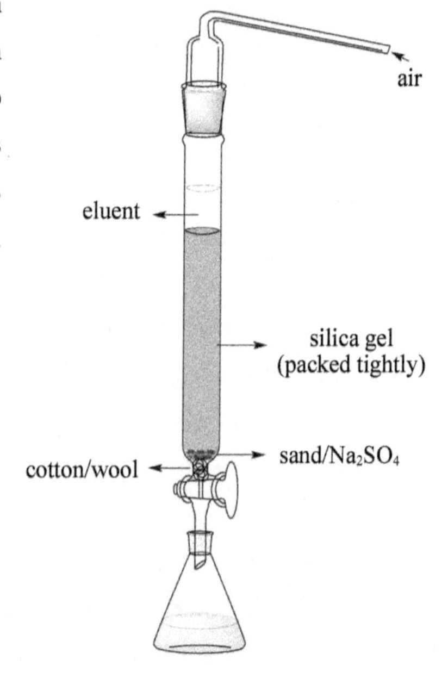

Figure 3.16 Flash chromatography

Columns are the apparatus used for flash chromatography. They are specific glass tubes that can hold the stationary phase, silica gel or aluminium oxide, and allow the elution of the mobile phase vertically. At the bottom of the glass tube, a Teflon stopcock is usually fitted with or without fritted glass bed while a 24/40 standard taper glass joint is at the top (Figure 3.16). Sometimes, ball-and-socket glass joints are used to replace the standard taper glass joints and columns without the fritted glass bed are more used because of smaller dead volume and lower chance for glass pieces freezing.

The stationary phase used for flash chromatography, silica gel or aluminium oxide, is usually of 230～400 mesh (38～63 μm). The larger ones of 70～230 mesh (63 ～ 219 μm) are used for gravity column without pressure applied. Solvents typically used for thin layer chromatography analysis can be used as the mobile phase in flash chromatography. Compressed air or inert gas such as nitrogen is usually applied to accelerate the separation and the flow rate of elute can be controlled by adjusting the gas pressures, but one should check the column for cracks or leakage before the operation and avoid too high pressure during the separation in that the pressure may force the detachment of the joints and cause accidents.

One wants to choose the size of a column based on the amount of the sample to be separated. This is to say that a small column should be used for a small amount of sample while a larger column is selected for a larger size of sample. The smaller R_f difference between the components in a sample usually requires a relatively larger size of column. A reference for column sizes in terms of R_f differences and sample amounts is briefly summarized in Table 3. 7.

Table 3. 7 Referenced column size and sample size①

Column's inner diameter (mm)	Sample size (g)		Size of each fraction (mL)
	$\Delta R_f \geq 0.2$	$\Delta R_f \geq 0.1 \sim 0.2$	
10	0. 1	0. 04	5
20	0. 4	0. 16	10
30	0. 9	0. 36	20
40	1. 6	0. 6	30
50	2. 5	1	50

The amount of the stationary phase also depends on the size of the sample to be separated. Usually, a height of 10 to 20 cm of stationary phase is adopted and a ratio of 8 : 1 to 10 : 1 the height of stationary phase to the inner diameter of the column is used. Typically, a shorter and fatter stationary column often provides poorer retention of components on the stationary phase but less diffusion and therefore always results in better separation. On the opposite, a taller and thinner column may hold the components longer in the stationary phase and polar solvents with higher elution strength are always required to pull the components off the stationary phase,

① W. Clark Still, Michael Kahn, Abhijit Mitra. Rapid chromatographic technique for preparative separations with moderate resolution, *Journal of Organic Chemistry*, 1978, 43(14), 2923 - 2925.

which would poorly discriminate between the compounds on the column because of molecular diffusion. Correspondingly, enough stationary phase must be used but using more stationary phase than necessary is not good for the efficiency of separation. One may follow a rough receipt that 1 g of sample uses ~15 g of silica gel, which corresponds to ~50 mL of volume as the density of the silica gel is roughly 0.3 $g \cdot mL^{-1}$. Then according to the column size as a diameter of 30 mm for ~0.9 g samples with a R_f difference higher than 0.2 as suggested in Table 3.7, the column height would be 7 cm. Or one can target with a column height of 15 cm to determine the column size with a diameter of ~1 cm in return.

As mentioned above, thin layer chromatogram is usually used to guide the choice of the elution solvents used in flash chromatography. The polarity of the commonly used elution solvents is increasing in this order: Alkane/cycloalkane <toluene< dichloromethane<diethyl ether<acetone<ethyl acetate<ethanol<methanol. One would use either the pure solvent or the mixture of two or several solvents as the elution liquid for flash chromatography. But the elution solvents should be as pure as possible because a small amount of polar impurities can dramatically change the polarity of the mobile phase and hence alter the elution behavior of the sample components. For instance, for the water miscible organic solvents such as acetone, ethanol, and methanol, the presence of water, which is much more polar than these solvents, can increase the elution strength significantly.

During the elution, fractions are collected. In order to avoid eluent containing different components that come off the column one after the other, smaller sizes of fractions are preferred as this can highly minimize the chance of re-mixing the components together. However, for a column to separate a sample of large sizes, a relatively high amount of eluent will be collected and too small size of each fraction would result in too many fractions. Therefore, in general, small amounts of the elution solvent is used for a small amount of sample and a smaller size of each fraction is collected. A large size of each fraction is collected for a larger column. This can also help to determine the size of collection containers, from several hundreds of microliters to tens of milliliters. Containers such as microcentrifuge tubes, test tubes, and Erlenmeyer flasks with appropriate sizes can be used for this purpose.

Each collected fraction is then subjected to thin layer chromatography analysis for identity and purity. The fractions with the same component(s), which is(are) desired for the experiment, should be combined and concentrated by rota-evaporation to afford the target. To minimize the work, one can analyze every three or four fractions depending on the number of fractions.

Technique summary for flash chromatography

When the size of column, the amount of silica gel, the elution solvents, and the fraction size are determined, one can follow the following steps to run a flash chromatography.

1. Pack a column (using silica gel as the stationary phase for instance)

a. The column without the fritted glass bed should be plugged at the stopcock using a small piece of glass wool or cotton in order to prevent the silica gel and sand from draining out.

b. A small amount of dried and clean sand (Anhydrous Na_2SO_4 can be used to replace sand.) is added to form a short layer of protection above the plug to hold the silica gel and prevent silica gel from draining out.

c. The stopcock is closed and the elution solvent is added to the column to a height of ~10 cm. Usually, the less polar solvent is used for this step if a solvent mixture is used as the elution solvent.

d. A pre-mixed slurry of silica gel in the desired amount (~1.5—2 times the volume of silica gel) of the elution solvent (or the less polar solvent) (the silica gel should be thoroughly and rigorously mixed with the solvent to remove all air present in the silica gel) is poured into the column. Then small amounts of the fresh elution solvent (or the less polar solvent, as used in the pre-mixed slurry) is used to rinse and flush the slurry in the original flask into the column and rinse down any silica gel stuck at the side of the column inner wall.

e. The stopcock is opened to allow the solvent to flow into a collection container and apply pressure to pack the column by compressed air or nitrogen until the solvent level is just into the top of the silica column and even with the height of the column. The top of the silica column should be flat. If not, more solvent should be added and the silica bed should be further stirred using a stir rod and then allowed to settle undisturbed with pressure applied. Never dry the column as air present in the column may influence the separation.

2. Load the sample onto the column

a. The sample is dissolved in minimal amounts of the solvent of low polarity (dichloromethane is frequently used because of its medium polarity and good dissolving capability).

b. The solution is carefully transferred to the column by dripping it down the inner wall of the glass column.

c. Rinse the original container with a small amount of the solvent and transfer it into the column the same way as mention in "b", and apply pressure to the column to push the solution into the column without drying the column.

d. Repeat "c" three times to achieve complete transfer of sample into the column.

e. Add some dry and clean sand to the column up to a height of ~1—2 cm.

3. Elute the column

a. Add the eluent solvent to the column slowly and carefully without disturbing the top of the column.

b. Apply pressure to the column and collect the eluent using test tubes or Erlenmeyer flask with appropriate sizes.

c. Change the collection containers once the current one is filled (try hard not to lose eluent during the container switching process).

d. Pay attention to the solvent in the column and replenish it as quickly as needed. Never dry the column.

e. The polarity of the elution solvent might be **slowly** increased for compounds with higher R_f values to come off quickly.

4. Identify and combine fractions

a. Samples in each fraction, or every several fractions are analyzed using thin layer chromatography to identify the compounds of different R_f values and the purity of the fractions.

b. Fractions with the same identify that is desired for the experiment and of similar purity are combined in a round bottom flask. The size of the round bottom flask is chosen based on the size of all the fractions that are combined. This analysis and combination should be undertaken while the column is still going in order to save time and get information on how the column goes.

5. Concentrate the fractions by rota-evaporation

The solution combined in step 4 is concentrated by rota-evaporation in vacuum.

6. Dry out the column

a. When the compounds to be collected are all out from the column, the flash chromatography is about to finish.

b. The air is allowed to push all the remaining solvent out from the column.

c. Continue the air flush until the silica gel is dry.

7. Dump the dried silica gel and sand into the solid waste container

Note:

1. As various volatile organic solvents are used in flash chromatography, the column should always be run in the fumehood.

2. The stationary phase used herein is tiny and light, and special safety precaution should be taken such as wearing a mask to prevent inhalation of the tiny particles into the lung when handling dry silica gels.

3. The packing method introduced in this part is called the *Slurry Method*. There is also a *Dry Adsorbent Method*, which directly load the dry stationary phase to the solvent-filled column, but it is no longer widely used as the stationary phase may fly to cause dust for potential safety issues.

4. Hexane, petroleum ether, and dichloromethane are often used to prepare the

silica slurry ahead of time.

5. Air bubbles, gaps or irregular surface present in the column may interrupt the separation.

6. The sample can be dissolved in solvents such as dichloromethane or methanol. Silica gel with larger meshes than that used for the flash chromatography is added to adsorb the sample molecules by removing the solvent through evaporation (such as rota-evaporation). The dried sample-adsorbed silica gel is poured to the pre-packed column as an alternative sample loading method. Such sample loading method is usually used for compounds that have poor solubilities in the solvent and require too much solvent by the solution loading method. When such sample loading method is used, precautions and protection must be applied to avoid inhalation of flying dust.

7. If necessary, additives, such as triethylamine or acetic acid, are added to the elution solvent for the flash chromatography.

8. Too high amount of the sample loaded on a column is called "*Overloading the column*" and may result in poor and incomplete separation.

9. Clamps or Clips are used to hold the joints together in order to attach the compressed air to the column.

10. A column should be run and completed as quickly as possible because diffusion occurs all the time. A lengthy operation of column or a column with a too slow elution rate can lead to severe diffusion and tailing, which can compromise the separation.

11. After rota-evaporation, the residue is usually further dried in vacuum to remove residual solvent for further use and analysis such as infrared spectroscopy, nuclear magnetic resonance spectroscopy, and mass spectrometry.

3.15.3 High Performance Liquid Chromatography

High performance liquid chromatography, short for HPLC, is one of the most powerful analytical methods based on separation used in the scientific field. With appropriate detection systems, it provides high sensitivity and excellent resolution for superior separation and analysis in comparison with other liquid chromatography methods. In general, HPLC requires a small amount of sample for analysis. Volatile and non-volatile samples can be analyzed. HPLC is powerful, but the instruments are usually of high cost and require some lab space to store.

There are several key components for an HPLC system, including solvent reservoir, (solvent degasser), pumps, injection system, column, detection system, instrument-computer communication system, and probably collection system. Most of the pieces are delicate and expensive while the column would be replaceable between different columns. The HPLC columns are usually commercial packed columns with a variety of functionalities and efficiencies. Among these, the reverse phase column might be the most frequently used in organic chemistry.

HPLC solvents can be any solvent with high purity while acetonitrile, methanol, and water are frequently used for the reverse-phase HPLC analysis. All solvents used in HPLC should be filtered through a 0.22 μm filter paper to remove any tiny solid particles. Samples should be prepared using the same eluent solvent and pass through a 0.22 μm～0.5 μm filter frit. For each injection, 10～100 μL of the solution is usually injected depending on the concentration of the solution in analytical HPLC while injection of higher volumes may be associated for preparative HPLC. One should not overload the column as this can result in poor separation and also saturate the detector for useless information.

The solvent eluent pattern can be programmed by the HPLC software and one needs to design the program based on the properties of the sample. Usually, isocratic program or gradient program is used for an HPLC analysis.

The important parameter one can obtain from the HPLC analysis is the retention time t_R, which is only dependent on the HPLC column, HPLC program, and the analyte. Therefore, it has a high reproducibility and is regarded as the characteristic data. The other is the peak height or peak area, which can demonstrate the purity of each sample and content of each component present in the sample.

3.15.4 Gas Chromatography

Gas chromatography, GC, is helpful to analyze volatile organic chemicals. It works with tiny amount of volatile compounds and relies only on the absorption capacity of the stationary phase with the analytes. Inert gases, such as helium or nitrogen, are used as the mobile phase and open tubular capillary columns are used to carry the stationary phase. As the mobile phase is gas, which is provided by the gas cylinder, no pump is required for any GC system while a flow regulator is used to control the flow rate of the gas. All chemicals within the column should be in the gaseous form and therefore the column is always at high temperature in an oven similar for the injection system and detection system. GC is more useful when in combination with a mass spectrometer as the detector for GC-MS analysis.

4 Application of Spectroscopy in Organic Laboratory

4.1 Introduction

It is important to determine the structure of the compound obtained from a reaction in order to verify if the original goal has been achieved, to collect information on how to optimize the reaction, and/or to acquire vital evidence in understanding the mechanism of the reaction. However, identification of organic compounds used to mainly rely on measuring their physical properties as melting point and boiling point, which provide limited information on the chemical component or molecular structure, until more valuable information has been provided by the recently developed spectroscopic methods, which have become routinely available in most research institutes. Four spectroscopic techniques, ultraviolet-visible spectroscopy (UV-Vis, section **4.2**), Fourier-transformed infrared spectroscopy (FT-IR, section **4.3**), mass spectrometry (MS, section **4.4**), and nuclear magnetic resonance spectroscopy (NMR, section **4.5**), are most powerful in determining structures of organic compounds on milligram scales in a matter of minutes. Because different spectroscopies can provide specific structural information in distinct aspects, these methods are usually used together to elucidate the structure of any organic compound. In general, the UV-Vis spectroscopy can reveal the presence of a conjugated π system or special chromophores[①]; The FT-IR spectroscopy identifies various functional groups; The MS can give information on the formula and possible atomic connectivity of the compound; The NMR discloses the local chemical and magnetic environment of a nucleus in addition to its relative quantity within the compound.

Instruments for UV-Vis and FT-IR spectroscopies are usually available in most institutes for the low cost and easy maintenance. Although NMR and MS instruments are very expensive and complex, the availability is increasing because of the enriched information provided for different compounds. For synthetic purposes, the frequency

① Any chemical group that absorbs light of certain wavelengths is defined as chromophore. Such groups are usually present in an organic compound, such as a dye or pigment, and are responsible for the interaction of the organic compound with light and for the apparent color of a compound.

of such techniques varies but falls in the general trend as NMR > FT-IR > MS > UV-Vis. This highly depends on the structural information that the technique can offer. For instance, NMR spectroscopy can provide a lot of useful information by both ^{1}H NMR and ^{13}C NMR in addition to the homonuclear or heteronuclear 2D NMR spectroscopies, while FT-IR can offer the information of various functional groups.

On the other hand, all the four spectroscopic methods use a small amount of sample for each test. The amounts of samples required for these methods are generally observed as following: MS < UV-Vis < FT-IR < NMR. Delightfully, all these spectroscopics techniques except MS are nondestructive, and therefore samples can be recovered after the measurement.

Among the four, UV-Vis, FT-IR, and NMR spectroscopies are absorption spectroscopy. The absorption spectrum shown in Figure 4. 1 is a plot of absorption or transmission of radiation against its frequency (ν) or wavelength (λ).

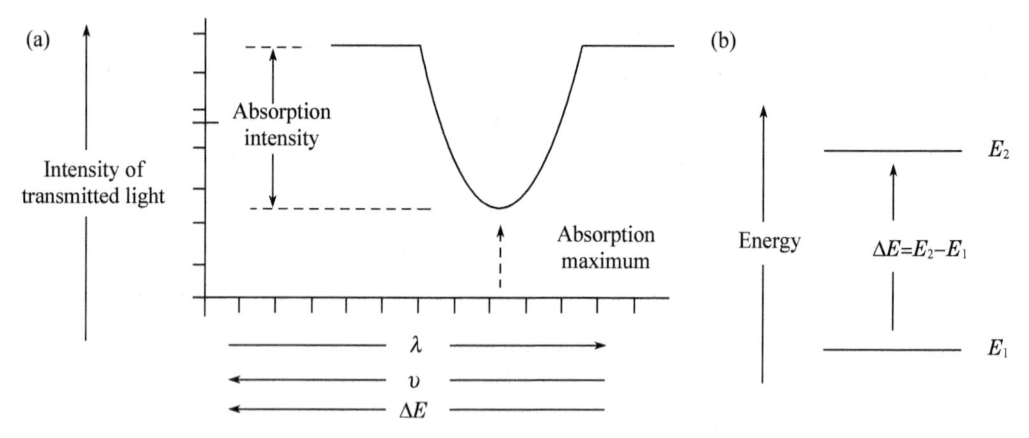

Figure 4. 1 Schematic absorption spectrum (a) and spectroscopic transition (b)

A molecule absorbs the energy (ΔE) and a spectroscopic transition occurs and takes the molecule from one state (E_1) to the other state with a higher energy (E_2). This energy change is given by:

$$\Delta E = E_2 - E_1 = h\nu$$
$$= h\frac{C}{\lambda}$$

where h is the Planck's constant and c is the speed of light in vacuum.

The frequency, wavelength, and energy of electromagnetic radiation are interrelated:

$$\Delta E \propto \nu$$
$$\Delta E \propto \frac{1}{\lambda}$$

A molecular spectrum, consisting of distinct bands or transitions correlated to the specifically quantized energy absorbed by the molecule, is characteristic and can be used to help to determine the structure of a molecule.

The MS is distinct from the others and relies mainly on the ionization of the compound and detection of the mass-to-charge ratio of the corresponding ions. It is used to establish the atomic connectivity and the molecular weight rapidly and efficiently.

4.2 Ultraviolet-Visible Spectroscopy

Ultraviolet-visible(UV-Vis) spectroscopy originates from electronic transitions within molecules upon the stimulation of electrons under the UV-Vis light irradiation. UV-Vis spectroscopy is a type of absorption spectroscopy. Most UV-Vis spectroscopy investigates the UV region of 200 to 400 nm and the visible region of 400 to 800 nm, in which the transitions of π-electrons or nonbonded electrons in the compound occur.

Take buta-1,3-diene for example, under irradiation with the UV light, it absorbs energy and a π electron is promoted from a bonding π molecular orbital to an antibonding π^* molecular orbital. It is the $\pi \rightarrow \pi^*$ excitation and UV light of 217 nm wavelength can stimulate such electronic transition, during which the energy of the photons matches well with the energy gap between the π and π^* orbitals shown in Figure 4. 2a.

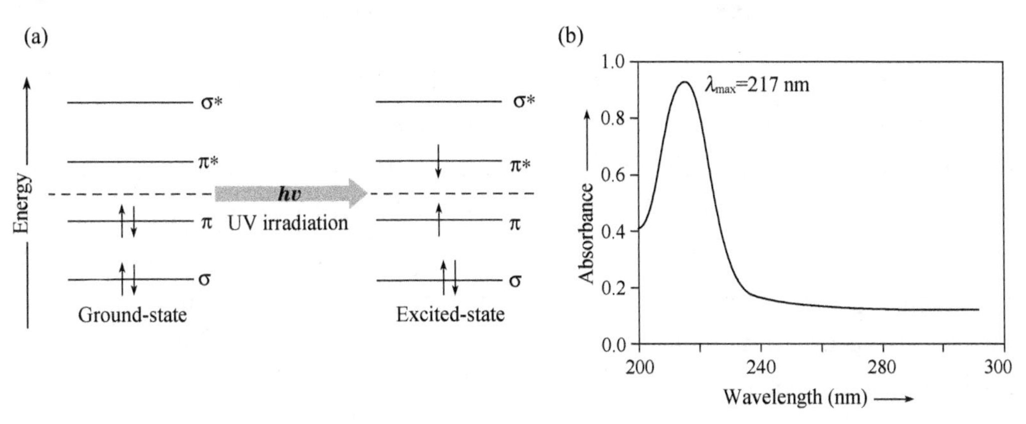

Figure 4. 2 (a) Ultraviolet excitation and (b) UV-Vis spectrum of buta-1,3-diene

Molecules in a number of different vibrational states undergo the same electronic transition, and it will only produce a band spectrum instead of a sharp one. Therefore compared with IR (section **4. 3**) and NMR (section **4. 5**) spectra, the UV-Vis spectra are featureless as in Figure 4. 2b.

Typically, UV-Vis absorption bands are characterized by the wavelength of the maximum absorption (the tip of the peak) labelled by λ_{max} and the strength of the absorption at such wavelength characterized by the molar extinction coefficient ε. For example, we describe a characteristic absorption of β-carotene as $\lambda_{max}=455$ nm ($\varepsilon=$ 23,000 L·mol^{-1}·cm^{-1}) in hexane as shown in Figure 4.3.

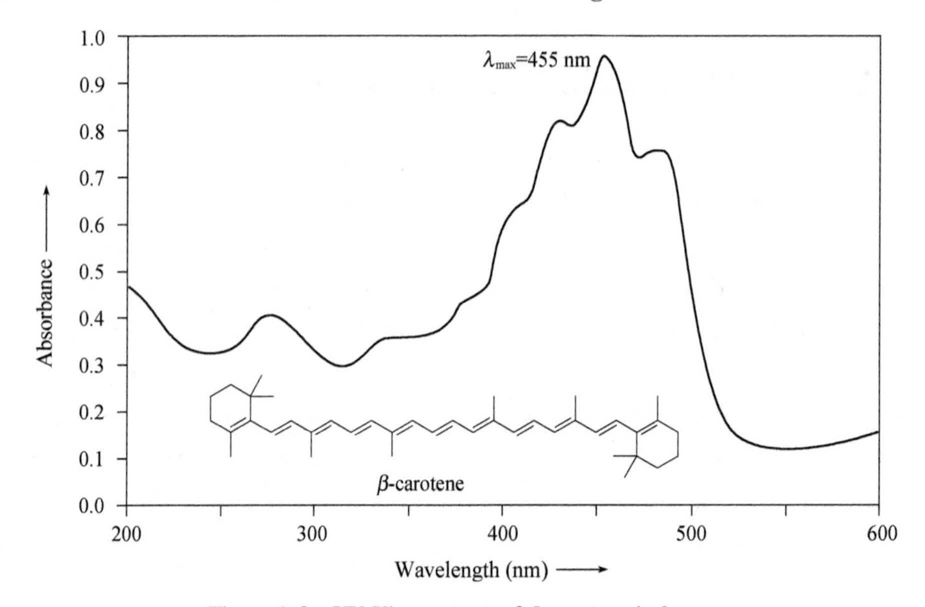

Figure 4.3 UV-Vis spectrum of β-carotene in hexane

The extinction coefficient ε is a fundamental and structure dependent constant. According to the Beer-Lambert Law, the UV-Vis absorbance A of a sample solution is defined as the logarithm of the ratio of the intensity of the incident light I_0 at a given wavelength to the transmitted intensity I. In addition, A is proportional to the solution's molar concentration c in mol·L^{-1} and the path length l in cm that the light travels through the sample solution. The equation shown below is usually used to calculate the concentration of a sample solution when the absorbance of the sample is measured and its molar extinction coefficient is known.

$$A=\lg\frac{I_0}{I}=\varepsilon cl$$

Aside from being used as a key detection method for high performance liquid chromatography (HPLC) or gas chromatography (GC), the UV-Vis spectroscopy is a precise tool for quantitative analysis of substances① such as highly conjugated organic compounds.

In the example of Figure 4.3, the absorbance at maximum wavelength of

① Detailed methods of quantitative analysis of sample concentration based on the Beer-Lambert Law including the use of calibration curves will be described in *Laboratory Experiments for Inorganic and Analytical Chemistry*.

455 nm, A, is 0.46, and the path length is 1.0 cm; So the concentration of the sample of β-carotene is 2.0×10^{-6} mol·L^{-1}.

$$c = \frac{A}{\varepsilon l} = \frac{0.46}{2.3 \times 10^{5} \frac{L}{mol \cdot cm} \times 1.0 \ cm} = 2.0 \times 10^{-6} mol \cdot L^{-1}$$

The wavelengths of absorption peaks are correlated with the types of functional groups in a given molecule and are valuable in determining the functional groups within a molecule. Ethylene has λ_{max} of 163 nm (ε=15,000 L·mol^{-1}·cm^{-1}) and buta-1,3-diene has λ_{max} of 217 nm (ε=20,900 L·mol^{-1}·cm^{-1}). As the conjugated system is extended, the wavelength of maximum absorption moves to longer wavelengths (toward the visible region). For instance, the maximum absorption of β-carotene, which consists of 11 conjugated double bonds, occurs at 455 nm.

Table 4.1 Rules to predict λ_{max} for conjugated polyenes and enones

Base value of parent (nm) / Functional groups / Increment (nm)	C=C—C=C (in hexane)* (1) Acyclic conjugated diene 217 (2) Heteroannular*** diene 214 (3) Homoannular**** diene 253 *****	C=C—C=O (in ethanol) 215**
Endocyclic double bond	36	39
Exocyclic double bond******	5	5
Double bond extending conjugation	30	30
Alkyl substituent or ring residue	5	α, 10; β, 12; γ, δ, 18
RCOO—	0	α, β, γ, δ, 6
RO—	6	α, 35; β, 30; γ, 17; δ, 31*******
RS—	30	β, 85
Cl—	5	α, 15; β, 12
Br—	5	α, 25; β, 30
—NR_2	60	β, 95

* Different solvent hardly affects the results.

** The calculation needs to include solvent correction (Table 4.2).

*** The two double bonds are located at two different rings.

**** The two double bonds are present in the same ring.

***** If both heteroannular diene and homoannular diene structures exist, use 253 nm as the base value.

****** Only for C=C bond, not for C=O bond.

******* These are for OCH_3.

Table 4. 2 Solvent Correction in predicting λ_{max} for conjugated polyenes and enones

Solvent	Correction factor to ethanol(nm)	Solvent	Correction factor to ethanol(nm)
Hexane	+11	Methanol	0
Ether	+7	Ethanol	0
Dioxane	+5	Water	−8
Chloroform	+1		

The Woodward-Fieser rules (Table 4. 1 and Table 4. 2), summarized from empirical observations, can be used to predict the maximum wavelength λ_{max} of conjugated organic molecules. For instance, the application of the rules in the above tables is demonstrated by the spectra of pulegone (**1**) and carvone (**2**) in hexane in Figure 4. 4.

Pulegone (**1**): $\lambda_{max}=215+10+12\times2+5-11=243$ nm(found 244 nm)

Carvone (**2**): $\lambda_{max}=215+5\times2+12-11=226$ nm (found 229 nm)

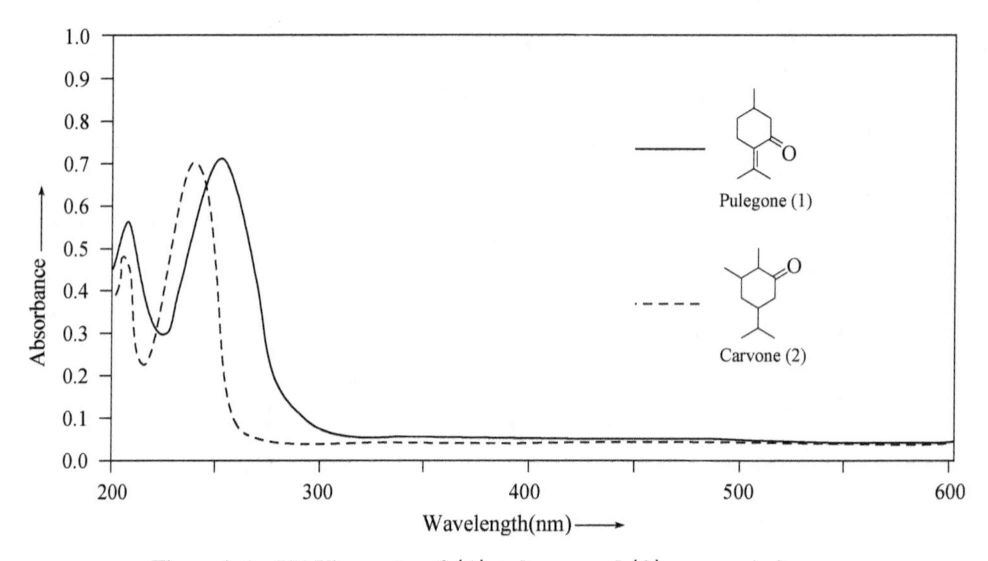

Figure 4. 4 UV-Vis spectra of (1) pulegone and (2) carvone in hexane

Compounds are dissolved in the solvent for the UV-Vis measurement. Various transparent solvents can be used, but the solubility, polarity, ionizability and cost of the solvents must be considered for an appropriate one in a test. In many cases, water is used for water soluble compounds while ethanol is chosen the most for organic compounds in that it has poor-to-no UV-Vis absorption at most wavelengths. When preparing the sample solutions, high concentrations are usually avoided to reduce the possibilities of molecule aggregation or exceeding the highest detection limit of the instrument.

Besides, cuvettes used in the measurements are made of quartz that absorbs no light. During the measurement, both the solvents and cells must be pure and clean. Since even a very tiny bit of an impurity or a fingerprint will result in UV-Vis light absorption, which would interfere the target spectrum.

1. *Instrumentation*

As schematically shown in Figure 4. 5, the basic parts of an UV spectrometer include a light source, a sample cell, a dispersing device (prism or grating), a detector, and a data output device.

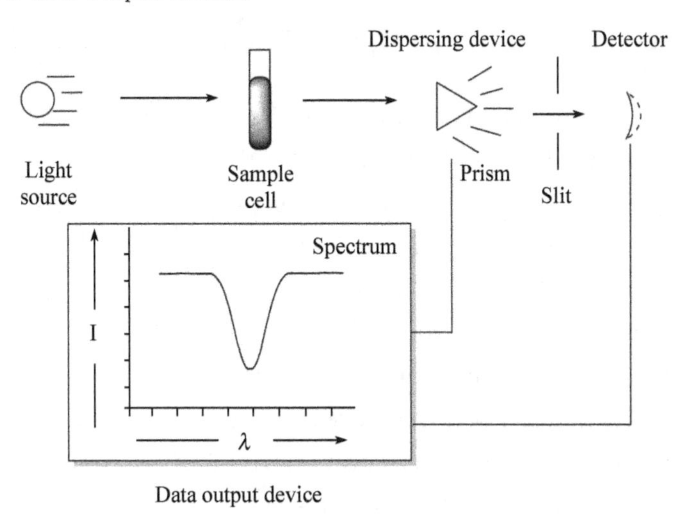

Figure 4. 5 Schematic representation of a UV spectrometer

The light source is often a deuterium arc lamp (185～395 nm) or a tungsten filament (350～800 nm), the combination of which can provide the whole range of 200 to 800 nm that is widely applied in the UV-Vis measurement.

In order to cancel out the absorption from the solvent as well as the atmosphere in the optical path, the UV-Vis spectrum of a blank sample is usually recorded before the test of the sample. Nowadays, double-beam instruments have been developed with an additional reference cell containing only the solvent for the same purpose.

2. *Experimental*

Determine an unknown compound (its concentration $c=5\times10^{-5}$ mol · L^{-1} in ethanol) provided by the instructor. The compound should be one of the following (The extinction coefficients of enones and polyenes are up to 10,000 L · mol^{-1} · cm^{-1}, so only very dilute solutions are needed.)

O　　　　O　　　　O

compound **A**　　compound **B**　　compound **C**

3. *Cleaning Up*

Since UV samples are extremely dilute solutions in ethanol, they can normally be collected in the non-halogenated waste solvent can.

4. *Question*

Predict the wavelengths of maximum absorption of the compounds **A**, **B** and **C** by the Woodward-Fieser rules.

4.3 Fourier Transform Infrared Spectroscopy

Infrared(IR) spectroscopy is another widely used molecular spectroscopy, which detects the absorption of the infrared light by the tested sample and records the spectrum based on the absorption or the transmission of the infrared beam. IR works in the wavelength range of 0.78 to 100 μm and involves molecular transitions among translational, rotational, and vibrational energy levels. The most useful IR range for organic compounds falls in a narrow range of 2.5 to 25 μm, which correlates to the mid-infrared wave number of approximately 4 000～400 cm^{-1} and associates with the fundamental vibrational and rotational modes. Since nearly all functional groups have characteristic and unique vibration frequencies, IR spectroscopy is the most used to identify the presence of such functional groups.

The vibrational and rotational transitions responsible for IR bands are quantized and dependent on the types of molecular vibrations as indicated in Figure 4.6, *i. e.* to periodic motions involving stretching or bending of bonds. It has been found that polar bonds are associated with strong IR absorption, while **SYMMETRICAL BONDS MAY NOT ABSORB AT ALL.**

Picture a bond in a molecule as a spring connecting two weights (atoms). The vibrational frequency of the bond should follow the classical Hooke's law. Therefore, shorter and stronger bonds have their stretching vibrations at the higher energy region (shorter wavelength) of the IR spectrum than the longer and weaker bonds. Similarly, bonds to lighter atoms (*e. g.* hydrogen), vibrate at higher energy than

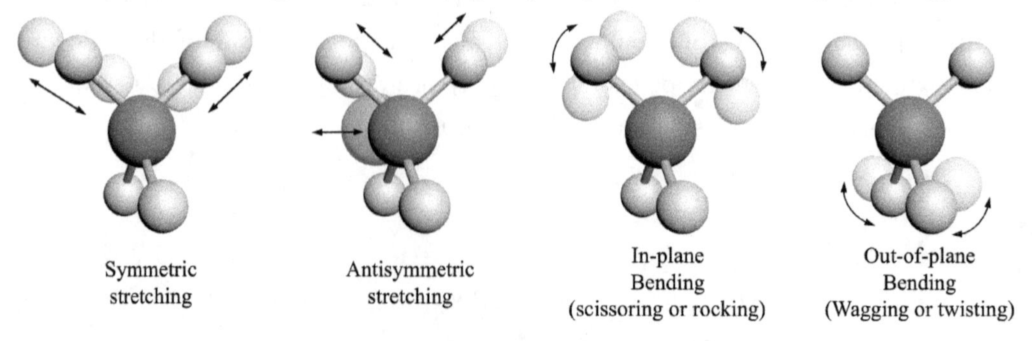

Figure 4.6 Modes of molecular vibrations

bonds to heavier atoms. Thus the IR absorption is determined almost only by the kinds of bonds and atom weights of a functional groups.

IR spectra are calibrated in wavelengths λ (μm) or in wavenumbers σ (cm^{-1}) with the relationship shown below:

$$\sigma\ (cm^{-1}) = \frac{1 \times 10^4}{\lambda\ (\mu m)}$$

Generally, wavenumber is more commonly used in IR spectra.

Table 4.3 Characteristic infrared absorption warenumbers

Compounds	Functional group	Wavenumber (cm^{-1})	Intensity*
Alcohol	O—H (free)	3 640—3 610	*s*, *br*
Alcohol	O—H (H bonded)	3 500—3 200	*s*, *br*
Carboxylic acid	O—H	3 100—2 500	*s*, *br*
Amine	N—H	3 500—3 300 (1° doublet, 2° singlet)	*m*
Terminal alkyne	*sp*-C—H	3 315—3 270	*s*
Olefinic and aromatic	sp^2-C—H	3 080—3 020	*m*
Aliphatic	sp^3-C—H	2 990—2 850	*m*
Aldehyde	C—H	2 900—2 700	*m*
Nitrile	C≡N	2 300—2 200	*m*
Terminal alkyne	—C≡C	2 260—2 210	*s*
Internal alkyne	—C≡C	2 140—2 100	*w*
Ester	C=O	1 750—1 740	*s*
Aldehyde or ketone	C=O	1 740—1 700	*s*
Amide	C=O	1 715—1 650	*s*
Unsaturated ketone	C=O	1 680—1 660	*s*
Alkene	C=C	1 675—1 640	*m*
Aromatic ring	C=C	2 000—1 660 1 600—1 450	*w* *m*
Nitro	NO_2	1 540	*s*
Aliphatic	sp^3-C—O	1 280—1 000	*s*
	sp^3-C—Cl	800—600	*s*
	sp^3-C—Br	600—500	*s*

* *s*, strong; *m*, medium; *w*, weak; *br*, broad.

Characteristic wavenumbers of certain functional groups have been listed in Table 4. 3. Noticeably, the electronic effect affects the frequencies: the resonance effect causes the frequencies to move to the lower wavenumber region, but the electron withdrawing effect does the opposite. ①

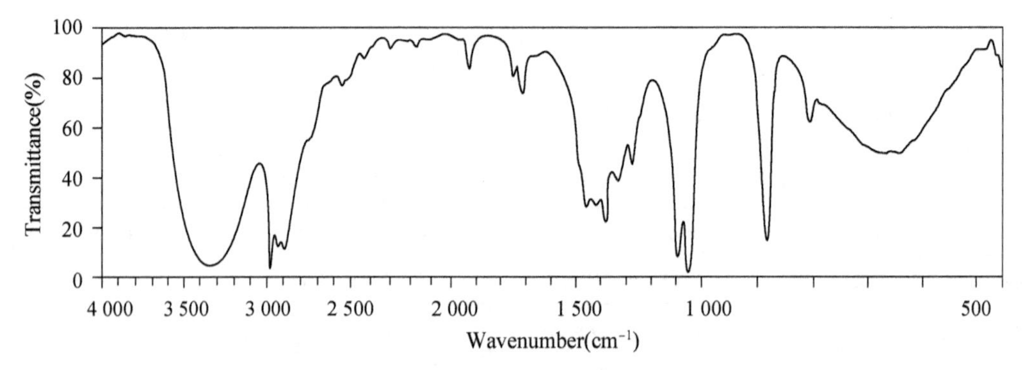

Figure 4. 7 An infrared spectrum of CH_3CH_2OH

Figure 4. 7 shows the IR spectrum of ethanol. Noticeably, the IR spectra are inverted and the x-axis, which is usually presented on the different scale, records the wavenumbers where the absorption occurs. A transmittance of 100% recorded in the y-axis means that all the infrared light passes through the sample, whereas a lower transmittance means that absorption occurs at certain frequencies as the IR light goes through the sample. Thus, each downward spike corresponds to an energy absorption. Peaks to the right of 1 500 cm^{-1} are complex and crowded. This part of the spectrum is called the "fingerprint region". Peaks to the left are sparse and this part is often referred to the "characteristic region", in which IR peaks for stretching vibrations of Y—H bonds, double bonds or triple bonds are usually located.

There are dozens of molecular vibrational motions even for compounds as simple as ethanol, which results in dozens of absorptions. So an IR absorption spectrum is much more complicated than a UV-Vis spectrum. On one hand, it limits the laboratory use of IR spectroscopy, only for pure samples of fairly small molecules. On the other hand, an IR spectrum acts as a unique fingerprint of a certain compound which makes it a powerful tool for compound identification. That means even two compounds have same boiling points or melting points or have identical UV-Vis spectra, they must be different in IR spectra.

Moreover, IR spectroscopy is rarely used for quantitative analysis. Thus, the intensities of IR peaks are neither proportional to concentration (unlike UV-Vis spectroscopy), nor to the numbers of atoms in the functional groups (unlike NMR

---

① For detailed influencing factors and rules, please refer to Ian Fleming, Dudley Williams, *Spectroscopic Methods in Organic Chemistry*, 7 ed., Springer International Publishing, 2019.

spectroscopy). Infrared absorption intensities are described by general classifications of *s* (strong), *m* (medium) or *w* (weak), and depend on the nature of molecular bonds.

Rules for IR spectrum analysis:

(1) Pay more attention to bands to the left (shorter wavelength) of 1,500 cm^{-1} and the strongest absorptions. Most functional groups have characteristic IR absorption bands in this area, which does not change from one compound to another (shown in Table 4.3). For example, the strong O—H absorption of a free alcohol is almost always in the range of 3 640 to 3 100 cm^{-1}; The strong C=O absorption of a ketone is usually in the range of 1680 to 1750 cm. With such characteristic absorption in hand, it is possible to identify the presence of functional groups from the IR spectra.

In the case of spectrum shown in Figure 4.7 for ethanol, in the area of the characteristic region, strong O—H absorption is in the range of 3 500 to 3 300 cm^{-1}; the medium absorption peaks of 2 960, 2 930, and 2 885 cm^{-1} indicate the existence of C—H stretching vibrations.

(2) The absence of characteristic peaks will definitely exclude certain functional groups. For instance, the absence of peaks in the area of 1 750 to 1 600 cm^{-1} (Figure 4.7) indicates that there is neither a C=C bond nor a C=O bond in the tested compound.

(3) Be aware of weak O—H peaks or the peak at 2 350 cm^{-1}, which indicates the possible presence of water① or CO_2 in the sample.

(4) If two samples have identical IR spectra, they are certainly identical in component.

(5) There are three most powerful IR spectra libraries, *Sadtler Infrared Spectra*, *Documentation of Molecular Spectroscopy* (*DMS*), and *the Aldrich Library of Infrared Spectra* available for pure compound or material identification.

1. *Instrumentation*

A simple dispersive IR spectrometer consists of the similar basic layout as that of UV-Vis spectrometer (shown in Figure 4.5). Nowadays, *Fourier Transform Infrared Spectroscopy* (FTIR) is more sophisticated because of its faster scanning speed, wider spectral range, higher sensitivity and resolution. It records an infrared interference pattern generated by a moving mirror, and the interferogram signal is converted by a computer using a mathematical technique called Fourier Transformation.

① Because KBr is hygroscopic, water signal is often found in the spectra of KBr pellets.

2. *IR sample preparation*

Samples in neat liquids, solutions in an appropriate solvent (10% solution in CS_2, CCl_4 or $CHCl_3$), or on solids as mulls and KBr pellets can be used for the IR measurement. However, the KBr disk method for both solid and neat liquid samples are commonly used in organic laboratories, while KBr is transparent over the whole IR range and raises no interruption for the measurement.

3. *A general procedure for the IR measurement using the KBr disk method*

(1) Make a KBr disk. KBr in spectroscopic grade should be used to prepare the IR sample. Since KBr is hygroscopic, it should be previously dried in an oven and stored in a desiccator. Similarly, the entire grinding process needs to be operated under an infrared lamp. About 1.5 mg of solid (pure compound) is added to 300 mg of spectroscopic grade KBr①, then gently and fully mixed and ground in an agate mortar until the mixture becomes a mull.

(2) Place the mull into a die that is subjected to a certain pressure by tightening the machine screws for several minutes or, under vacuum in a specially constructed hydraulic press. The most simple, small press illustrated in Figure 4.8 consists of a large nut and two machine screws. The sample is placed between the two machine screws, which have polished surfaces. After being tightened, the screws are then loosened and removed, with the KBr disk left in the nut. Pressure and time determine the quality of the newly formed KBr disk. The disk should be transparent. An opaque area in the disk indicates that insufficient pressure was applied. Too much pressure can result in crushing the disks. For a neat liquid sample, place a drop of the liquid onto the freshly prepared KBr disk before being mounted in the spectrometer.

(3) Remove the disk from the die with tweezers and placed in a special holder in the sample cell of the IR spectrometer.

(4) Run the IR experiment and record the spectrum.

(5) Clean Up. The disks should be cleaned by rinsing them with an organic solvent such as acetone or ethanol and wiping them dry with a paper towel. Discard halogenated liquids in the halogenated organic waste container. Other solutions should be placed in the regular organic solvent waste container.

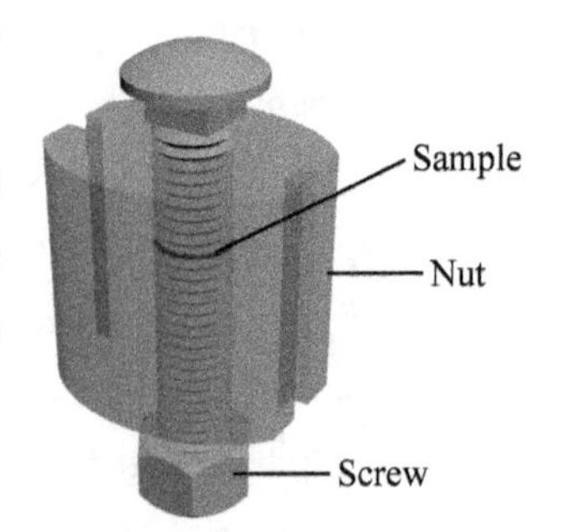

Figure 4.8 KBr disk die

4. *Experimental*

There are two unknown white powders **A** and **B**, one is benzoic acid, and the other is acetanilide. Try to

① To avoid that IR peaks are too weak or too intense, one should adjust the sample concentration for a proper spectrum. Generally, the most intense peak with an absorbance of about 1.0 is the best.

use IR spectroscopy to determine which is benzoic acid and which is acetanilide.

4.4 Mass Spectrometry

Mass spectrometry measures individual ion's mass-to-charge ratio (m/z) to provide information of the compound. Since multi-charged ions are much less abundant than those with a single electronic charge ($z = 1$), m/z is usually equal to the formula mass of the ion, m. Specific mass spectrometric technique is usually applied to generate ionic fragments from the source compound and measure the mass-to-charge ratios of the fragments to collect information on some atom connectivity in addition to the molecular weight.

1. *Instrumentation*

Generally, there are more than 20 kinds of mass spectrometers commercially available for different applications and they share the same basic parts. A mass spectrometer usually includes three basic parts: An *ionization source* to convert the sample molecules into ionic forms, a *mass analyzer* to sort ions based on their mass-to-charge ratios, and a *detector* to observe and count the separated ions, as shown in Figure 4.9.

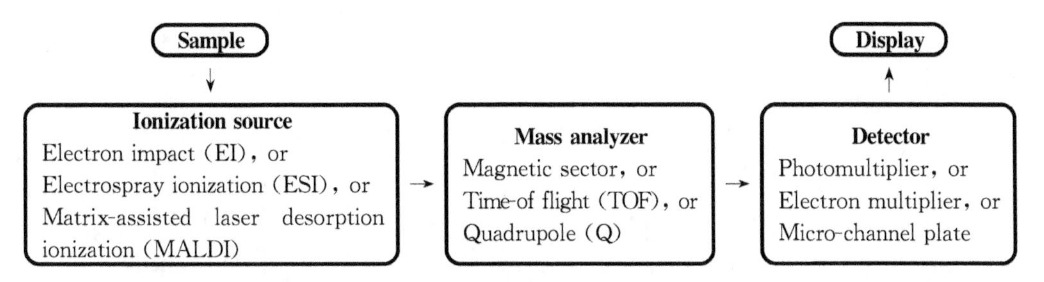

Figure 4.9 Composition of an electron-ionization, magnetic-sector mass spectrometer

One of the most commonly used mass spectrometers is the electron-impact, magnetic-sector instrument, and it is most suitable for volatile compounds and commonly used in combination with GC in the organic laboratory. Usually, a small amount of sample is vaporized into the ionization source under a high vacuum, where it is bombarded by a stream of energetic electrons (70 eV). One high energetic electron e collides into a molecule M to dislodge a valence electron from the molecule, producing a *cationic radical*. And the cationic radical falls apart into smaller pieces. Some pieces retaining positively charged will be deflected into different paths because of different mass-to-charge ratios and sorted onto the detector when they flow through a magnetic field.

$$e^- + M \longrightarrow M^{\dot{+}} + 2e^-$$

Organic molecule   *Cationic radical*

As shown in Figure 4.10, the mass spectrum of a compound is typically presented as a bar graph, with relative abundance of ions striking the detector on the y axis and masses (m/z values) on the x axis. The strongest peak, assigned with an intensity of 100%, is called the *base peak* (B), and the peak that corresponds to the unfragmented cationic radical is called the *parent peak*, or the *molecular ion* ($M^{\dot{+}}$). Fragmentation patterns of the mass spectrum are usually complex and quite informative, while the molecular ion is not necessarily the base peak. The mass spectrum of acetophenone in Figure 4.10 shows that the parent peak at m/z of 120 is only about 30% as high as the B peak at m/z of 105. Besides, many other fragment ion peaks are present.

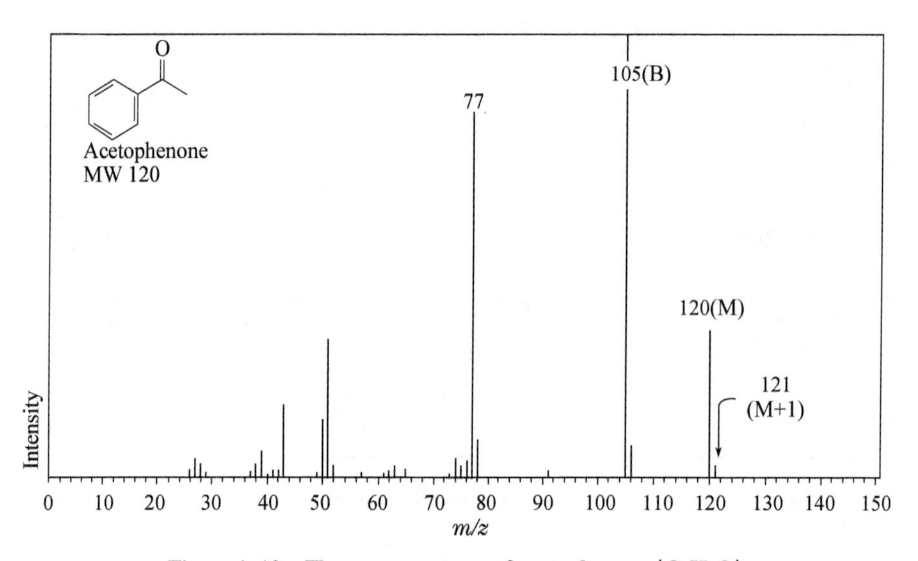

Figure 4.10 The mass spectrum of acetophenone (C_8H_8O)

2. *Rules*

(1) The m/z value of the molecular ion is equal to the molecular weight (MW) of the compound. Noticeably, in addition to the molecular weight, the cracking pattern also provides useful structural information to reconstruct the intact precursor. In an organic laboratory, with the knowledge of the chemicals used in the reaction and the sample (unknown product), it's useful to deduce the molecular formula and even speculate the structure of the product. For example, treatment of benzoic acid with ethanol in presence of concentrated H_2SO_4 gives an oily product with a molecular ion at 150 (m/z). This molecular ion corresponds to the addition of 28 mass units to the

starting material, consistent with the formation of an ethyl ester by condensation of the carboxylic acid and the alcohol and formation of a water molecule.

OH (MW=122) + CH_3CH_2OH (MW=46) $\xrightarrow[\text{reflux}]{\text{conc. } H_2SO_4}$ ($M^{+\bullet}$ m/z=150) ⇒ O

Table 4.4 Isotope information for some elements frequently present in organic compounds

Element	Isotope	Natural abundance (%)	Exact mass (amu)
Hydrogen	^{1}H	99.99	1.007 83
	^{2}H	0.01	2.014 10
Carbon	^{12}C	98.93	12.000 0 (standard)
	^{13}C	1.07	13.003 4
Nitrogen	^{14}N	99.63	14.003 1
	^{15}N	0.37	15.000 1
Oxygen	^{16}O	99.76	15.994 9
	^{17}O	0.04	16.999 1
	^{18}O	0.20	17.999 2
Fluorine	^{19}F	100	18.998 4
Silicon	^{28}Si	92.23	27.976 9
	^{29}Si	4.68	28.976 5
	^{30}Si	3.09	29.973 8
Phosphorus	^{31}P	100	30.973 8
Sulfur	^{32}S	94.93	31.972 1
	^{33}S	0.76	32.971 5
	^{34}S	4.29	33.967 9
	^{36}S	0.02	35.967 1
Chlorine	^{35}Cl	75.78	34.968 9
	^{37}Cl	24.22	36.965 9
Bromine	^{79}Br	50.69	78.981 3
	^{81}Br	49.31	80.916 3
Iodine	^{127}I	100	126.904 5

(2) Isotopes of each elements in the compound may contribute to the presence of different individual molecules and a unique parent molecular mass is available for each molecule. Correspondingly, there are a series of mass-to-charge ratios available for each fragment because of the presence of isotopes. This is usually regarded as the isotope distribution. For the simplest organic compound CH_4, because of the presence of ^{12}C and ^{13}C as well as ^{1}H and ^{2}H naturally as shown in Table 4. 4, there are several possible atom combinations as $^{12}C^{1}H_4$, $^{12}C^{1}H_3{}^{2}H$, $^{12}C^{1}H_2{}^{2}H_2$, $^{12}C^{1}H^{2}H_3$, $^{12}C^{2}H_4$, $^{13}C^{1}H_4$, $^{13}C^{1}H_3{}^{2}H$, $^{13}C^{1}H_2{}^{2}H_2$, $^{13}C^{1}H^{2}H_3$, and $^{13}C^{2}H_4$. This would afford more than one mass-to-charge ratio for the parent molecular ions and several peaks with distinct intensities would be viewed in its mass spectrum. The analysis of isotopic peaks of the molecular ion can be used to infer the presence of special elements based on their natural isotopic abundance. In nature, the isotopes ^{79}Br and ^{81}Br occur in a ratio of 51 : 49, while the isotopes ^{35}Cl and ^{37}Cl occur in a ratio of 76 : 24. Thus the intensity of the peak at $M+2$ (m/z = 114) in chlorobenzene must carry the isotopic abundance information of chlorine-37 and is about 1/3 of that of the peak at M (m/z = 112) which equals the relative abundance ratio of the two chlorine isotopes as in Figure 4. 11a, and the intensity of the peak at $M+2$ (m/z = 168) for ethyl bromoacetate is almost the same as that of the peak at M (m/z = 166) which is attributed to the relatively close abundance of the two bromine isotopes shown in Figure 4. 11b.

(3) High-resolution mass spectrometry (HRMS) is a highly precise mass measurement which can measure m/z values accurately to 5 ppm (four decimal places) and are capable of determining the molecular formula of the molecular ions by measuring the exact mass of the compound. Note, the exact mass is not the molecular weight (the sum of the average atomic masses of elements on the periodic table) but the sum of the exact atomic masses of the specific isotopes, which means that its measurements should refer to molecules with specific isotopic compositions. The exact masses of some elements and their most abundant isotopes are given in Table 4. 4. For example, in the low-resolution mass spectrum, all of these compounds $C_6H_{12}N_2$, $C_7H_{14}N$, and $C_7H_{12}O$ give the mass-to-charge ratio of 112 (m/z) for the molecular ions. Using the data in Table 4. 4, the molecular exact masses are calculated to be:

$C_6H_{12}N_2$: $(6\times12.0000) + (12\times1.00783) + (2\times14.0031) = 112.1000$

$C_7H_{14}N$: $(7\times12.0000) + (14\times1.00783) + (1\times14.0031) = 112.1126$

$CH_{12}O$: $(7\times12.0000) + (12\times1.00783) + (1\times15.9949) = 112.0888$

On the one hand, if the exact mass can be determined to 0. 001 amu, it's obviously easy to distinguish these three compounds. For instance, if the measured

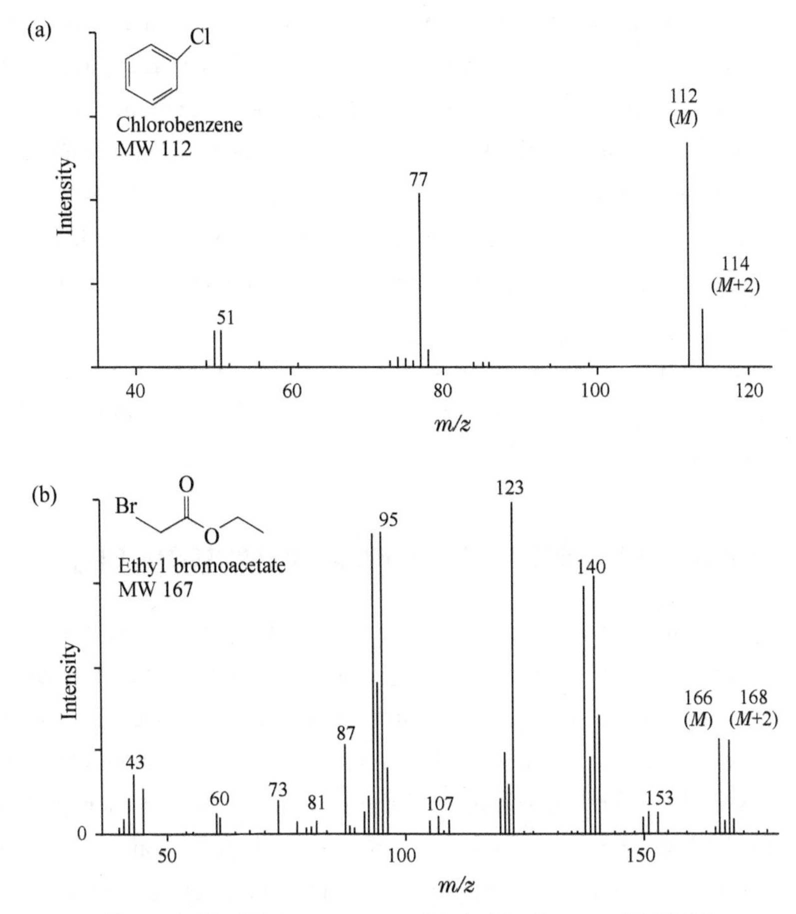

Figure 4.11 The mass spectra of (a) chlorobenzene (C_6H_5Cl) and (b) ethyl bromoacetate ($C_4H_7O_2Br$)

exact mass is 112.112 4, the molecule formula $C_7H_{14}N$ can be given by computer programs developed for commercial mass spectrometers.

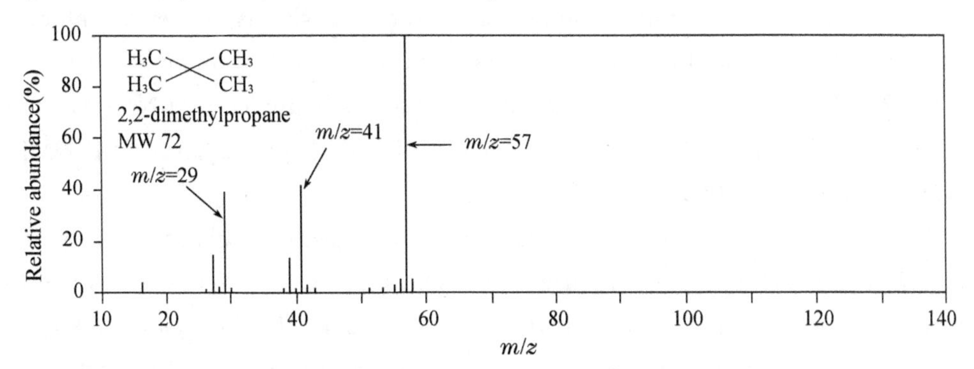

Figure 4.12 The mass spectrum of 2,2-dimethylpropane (C_5H_{12})

(4) Unfortunately, due to the high energy of EI ionization source, some fragile compounds would only show fragmented ion peaks. As shown in Figure 4.12, there is no molecular ion in EI mass spectrum of 2, 2-dimethylpropane. In these cases, alternative "soft" ionization methods are applied, *i. e.* electrospray ionization (ESI) or matrix-assisted laser desorption ionization (MALDI) that do not use electron bombardment can prevent or minimize fragmentation. Especially, linked to a time-of-flight mass analyzer, it can analyze biochemicals, even those with very high molecular weight.

3. Experimental

The product from Experiment **5.6** (Synthesis of Ethyl Acetate) is dissolved in ethanol and analyzed by EI.

4.5 Nuclear Magnetic Resonance Spectroscopy

When atomic nuclei with non-zero nuclear spin quantum number are placed in a magnetic field, the nuclei will behave like magnet bars and orient dynamically in various ways and eventually reach an equilibrium. Once a radiofrequency electromagnetic field is applied, the nuclei whose frequency matches the radiofrequency will absorb the energy and the magnets at lower energy state will be stimulated to the higher energy state and a new spin distribution equilibrium will be established. Such phenomenon is usually referred as the nuclear magnetic resonance, short as NMR. Upon the removal of the radiofrequency electromagnetic field, the decaying signal is measured along the time course and a spectrum can be constructed through the Fourier transform method. Such technique, referred as nuclear magnetic resonance spectroscopy, is the most powerful tool in organic chemistry and can be used to determine the number, kind, and relative positions of certain atoms in molecules. By applying different types of spin nuclei in such technique, a variety of 1D and 2D NMR spectroscopies are developed, such as hydrogen (^{1}H NMR), carbon (^{13}C NMR), fluorine (^{19}F NMR), nitrogen (^{15}N NMR), phosphorus (^{31}P NMR), COSY, NOESY, HMBC, and HMQC①, *etc.*

Due to the presence of protons, any nucleus is positively charged. A spinning charged nucleus acting like a tiny magnet generates a magnetic dipole, gives rise to a magnetic moment, and is associated with a magnetic field *H*. *Spin quantum numbers* designated by the letter *I* are used to characterize the nuclear magnetic dipoles. The

① COSY is Short for Correlation Spectroscopy; NOSEY is short for Nuclear Overhauser Effect Spectroscopy; HMBC is short for Heteronuclear Multiple Bond Correlation Spectroscopy; HMQC is short for Heteronuclear Multiple-Quantum Coherence Spectroscopy.

spin quantum numbers I can be 0, 1/2, 1, 3/2, *etc.* Nuclei such as ^{12}C, ^{16}O, and ^{32}S with spin quantum numbers of 0 have no net spin. Other nuclei such as ^{1}H, ^{13}C, ^{15}N, ^{17}O, ^{19}F, and ^{31}P have finite magnetic moments with spin quantum numbers of $I = 1/2$ and are 'NMR visible'. So they are the most used in NMR measurements.

In particularly, hydrogen and carbon are the most common nuclei found in organic compounds, so that the most used NMR spectroscopies are ^{1}H NMR and ^{13}C NMR. The behavior of ^{1}H nuclei in magnetic fields will be taken as the model for other nuclei with spin quantum number $I=½$ in this section.

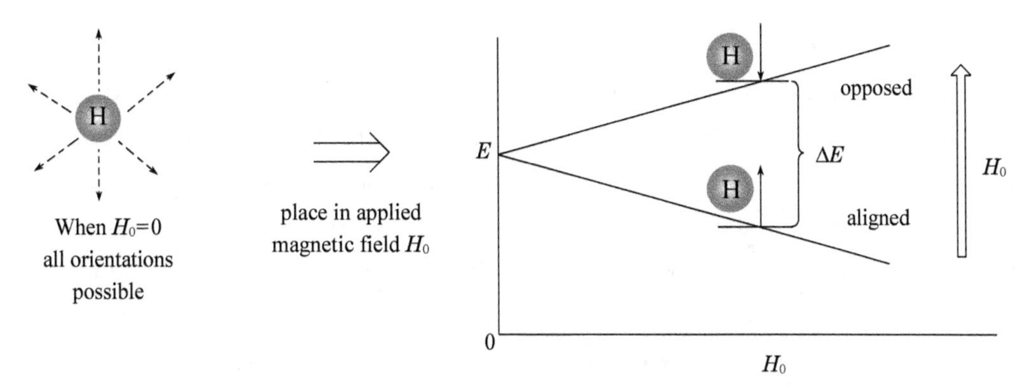

Figure 4. 13 Energy of the spin states of a hydrogen nuclei in a magnetic field

As shown in Figure 4. 13, after being placed in a uniform external magnetic field H_0, the nuclear magnetic moment adopts $(2I + 1)$ orientations. Otherwise, the spins of magnetic nuclei are oriented randomly. So the spin quantum number of $I = 1/2$ for the hydrogen nucleus results in only two permissible orientations of the nuclear moment relative to the direction of the applied field: Either aligned with H_0 (lower energy) or opposed to it (higher energy). It will induce a spectroscopic transition (spin-flip) by the absorption of a quantum of electromagnetic energy (ΔE) of the appropriate frequency (ν) if an additional magnetic field is applied. Since the magnetic moment of the hydrogen (γ), also called as magnetogyric ratio, is the same for all hydrogen nuclei and the energy difference between the two states is given by the following equation, where ΔE as well as ν is proportional only to the strength of the applied external magnetic field H_0 according to the Larmor equation:

$$\Delta E = h\nu = \gamma h H_0 / 2\pi$$

Unlike other spectroscopies, for every value of H_0, there is a matching value of ν in NMR corresponding to the condition of *resonance*. Therefore, as shown in Table 4. 5, *resonance frequencies* commonly found in commercial instruments reflect magnitudes of applied magnetic field H_0.

Table 4.5 Resonance frequencies of ^1H and ^{13}C Nuclei in magnetic fields

$\nu(^1H$, MHz)	$\nu(^{13}C$, MHz)	H_0 (Tesla)
90	22.6	2.113 9
300	75.4	7.046 2
400	100.6	9.395 0
500	125.7	11.744
600	150.9	14.092 3
900	226.3	21.128

If all ^1H nuclei in a molecule had absorbed energy at the same frequency, we would have only observed one single NMR absorption band in the ^1H spectrum of a molecule for a given magnetic field H_0. That would be useless. Thanks to the fact that all nuclei in molecules are surrounded by electron clouds. The electrons moving around protons induce tiny local magnetic fields H_{ind} of their own, acting to counter the applied magnetic field H_0. So that the effective magnetic field H_{eff} felt by the proton is different from H_0. We call this as the *shielding effect*:

$$H_{eff} = H_0 - H_{ind}$$

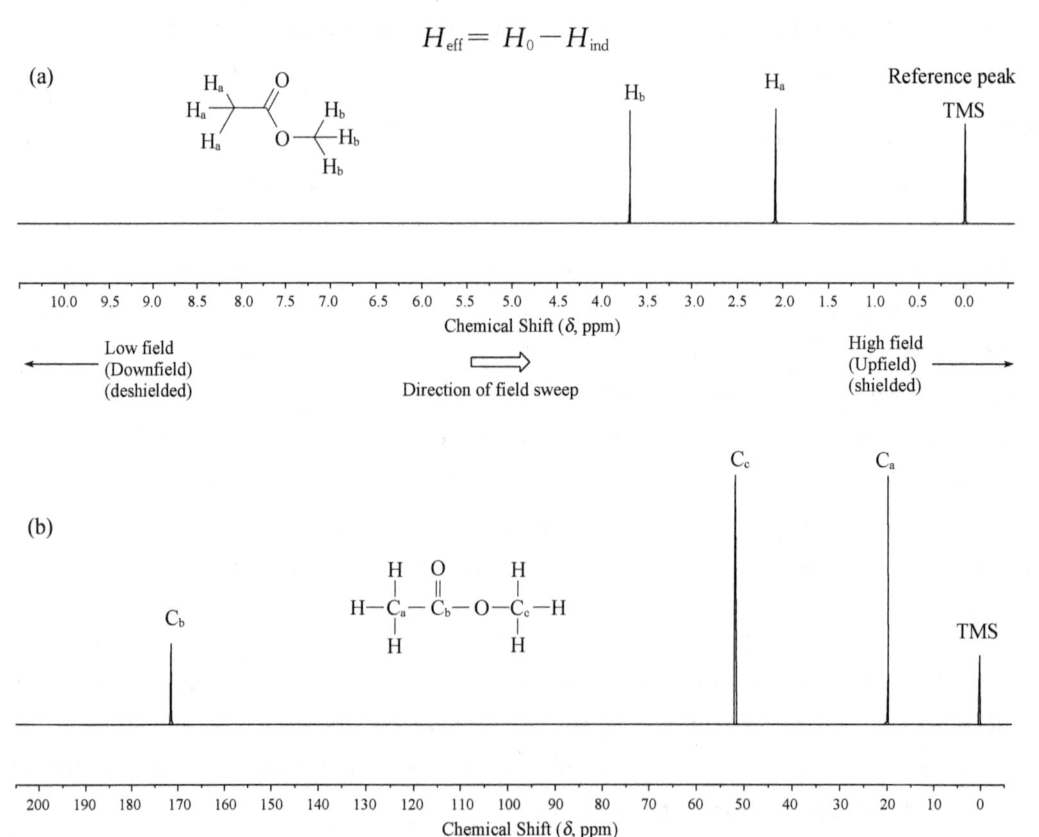

Figure 4.14 ^1H NMR (a) and ^{13}C NMR (b) spectra of methyl acetate

Because each chemically distinct proton in a molecule is in a slightly different electronic environment, the effective magnetic field felt by each nucleus is unique and slightly different. These tiny differences can be detected by ^{1}H NMR. Figure 4. 14a is the ^{1}H NMR spectrum of methyl acetate, which reveals the existence of two different types of protons (H_a and H_b) and these two types of protons resonate at two different frequencies. The left part of the chart represents the resonance of the nuclei at lower field (or called as downfield) while the right part correlates with the interaction at higher field (or referred as upfield). Nuclei absorbing on the downfield side of the chart require a lower radiofrequency field strength for resonance, implying that they have less shielding; While nuclei absorbing on the upfield side require a higher field strength for resonance, implying that they have more shielding. The horizontal axis *chemical shift* (δ) shows the effective field strength felt by the proton. The vertical axis indicates the intensity of absorption of energy. The small peak labeled "TMS" at the right of the spectrum is a reference peak. Same as the ^{13}C NMR spectrum shown in Figure 4. 14b, different types of carbons (C_a, C_b and C_c) are distinguished by different frequencies, at which they resonate, and result in three different peaks.

Chemical shift is defined as the position on the NMR spectrum at which a signal correlates with a nucleus spinning , so it's a frequency scale. A small amount of tetramethylsilane [$(CH_3)_4Si$, TMS] is used as a reference to calibrate the position of an absorption because it produces a single peak both in ^{1}H NMR and ^{13}C NMR that occurs upfield. The chemical shift of TMS is intentionally set as the zero point, and most of other absorptions occur to the left on the chart (downfield). An arbitrary scale (δ)① is used to calibrate the chart in dimensionless units called "parts per million" (abbreviated to ppm) with the following equation:

$$\delta(\text{ppm})=\frac{\text{frequency referred to TMS (Hz)}}{\text{Spectrometer operating frequency (MHz)}}$$

The chemical shift of a NMR absorption in δ units is constant, ranging from 0 to 15 ppm for most protons and from 0 to 220 ppm for most carbon atoms, regardless of the operating frequency of the spectrometer. Typical ^{1}H and ^{13}C chemical shift ranges with different electronic environment in organic compounds are shown in Figure 4. 15. In general, nuclei that are more strongly shielded by electrons require a higher applied field to bring them into resonance and therefore absorb on the right side of the NMR chart. Nuclei that are less shielded need a lower applied field for resonance and therefore absorb on the left of the NMR chart. The distribution of electronic clouds is

① By using a system of measurement in which NMR absorptions are expressed in relative terms (ppm relative to spectrometer operating frequency) rather than absolute terms (Hz), it's possible to compare spectra obtained on different instruments.

mainly affected by hybridization and electronegativity of atoms attached to the proton and other electronic effects. ①

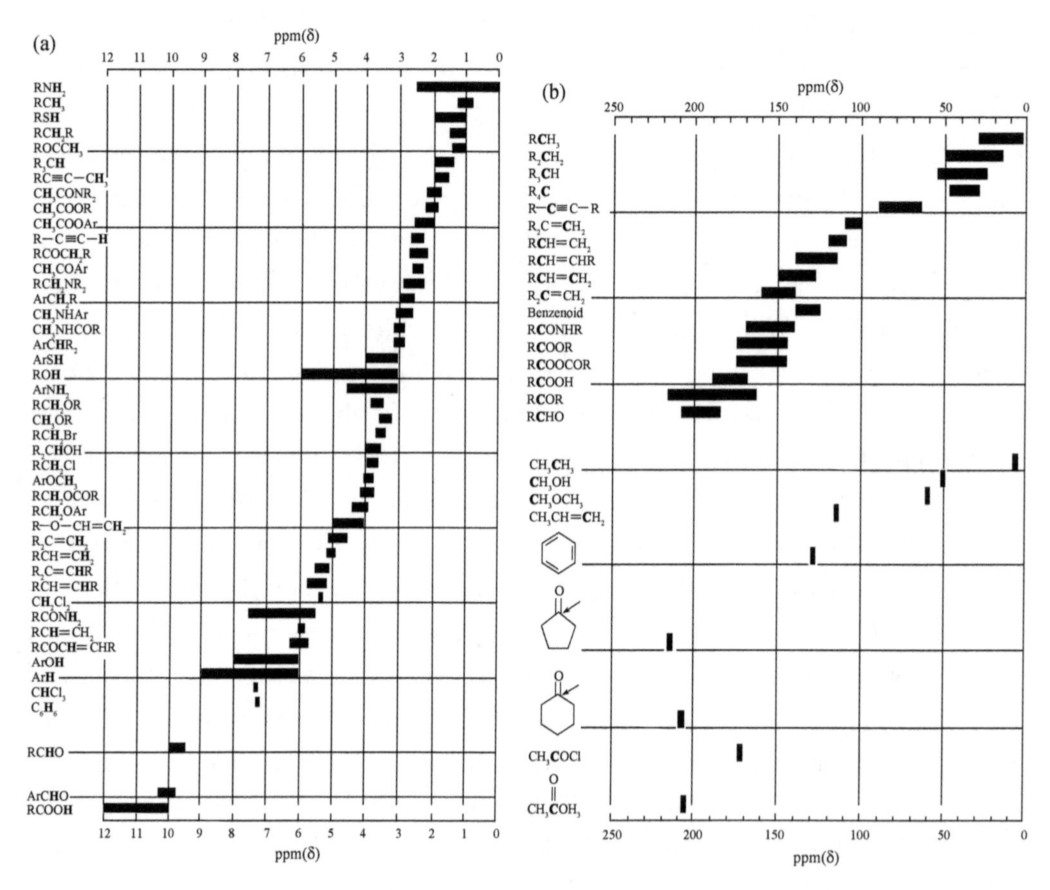

Figure 4.15 Typical (a) ^{1}H and (b) ^{13}C chemical shift ranges in organic compounds

Furthermore, ^{1}H NMR is not only able to distinguish different types of protons in a molecule, but also tells how many each type of protons there are (at least by ratio) by *integration* of the absorption intensity area of each peak. As shown in Figure 4.16, measuring the integrated area of each hydrogen peak of CH_3CH_2Br gives a 3 : 2 ratio, which corresponds to the ratio of the two types of protons.

Another exciting phenomenon called *spin-spin splitting* or *spin-spin coupling* is that the absorption of each proton splits into multiple peaks, called a *multiplet*, as shown in Figure 4.16. As each spin nucleus can be regarded as a bar magnetic, it can produce its own local magnetic field, which will affect the net magnetic environment of its neighboring spin nucleus. Such influence is generally determined by the nature and number of the spin nuclei and leads to multiple resonance peaks of a proton

① For detailed influencing factors and rules, please refer to Ian Fleming, Dudley Williams, *Spectroscopic Methods in Organic Chemistry*, 7 ed., Springer International Publishing, 2019.

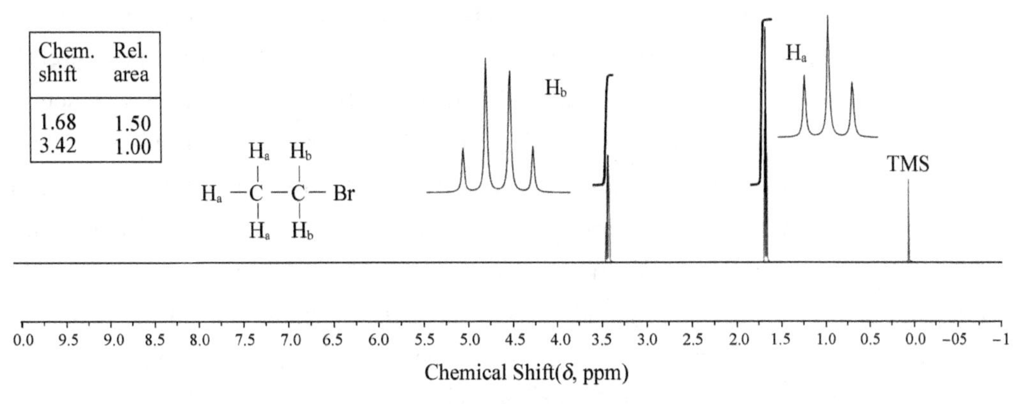

Figure 4.16 ^{1}H NMR spectrum of bromoethane

instead of a single one. The spin-spin coupling phenomenon is complex and the most general guidance is the $n+1$ rule. There are $(n+1)$ peaks in the NMR spectrum for protons that have n equivalent neighboring protons. For example, in Figure 4.16 the H_a($C\underline{H}_3$) signals of bromoethane appear as three peaks (a *triplet*) centered at 1.68 ppm and the H_b($C\underline{H}_2Br$) signals appear as four peaks (a *quartet*) centered at 3.42 ppm. The triplet is caused by splitting of the H_a signal with two equivalent neighboring protons on the methylene group ($n = 2$ leads to $2+1 = 3$ peaks). The quartet is due to H_b signal split by the three equivalent methyl protons ($n = 3$ leads to $3+1=4$ peaks). Integration confirms the expected ratio of 3 : 2. The intensity of the *spin-spin splitting* is usually characterized by the *coupling constant* (J) which is measured in hertz (Hz) and generally falls in the range of 0 to 18 Hz. The exact value of the coupling constant between two neighboring protons depends on the geometry of the molecule, and the same coupling constant is shared by both the coupled groups of hydrogens and is independent of the applied field strength. In bromoethane, for instance, the H_a($C\underline{H}_3$) is coupled to H_b($C\underline{H}_2Br$) and appears as a quartet with $J = 7$ Hz. The H_b($C\underline{H}_2Br$) appears as a triplet with the same coupling constant $J = 7$ Hz. Thus it is possible for a ^{1}H NMR spectrum to give information not only about numbers and kinds of protons in a molecule, but also the connectivity of the carbons bearing those protons. In other words, it is possible to tell which multiplets even in a very complex NMR spectrum are related to each other or which groups or carbon skeletons may be adjacent.

1. *Instrumentation*

For any nucleus the condition of resonance may be achieved by keeping the field constant and changing (or sweeping) the frequency or, alternatively, by keeping the frequency constant and sweeping the field. A schematic diagram of a field sweep NMR spectrometer is demonstrated in Figure 4.17. A NMR tube containing the sample solution is placed between the poles of a magnet and irradiated with radiofrequency

(RF) energy. If the frequency of the irradiation is held constant and the strength of the applied magnetic field is varied, each nucleus comes into resonance at a slightly different field strength. A sensitive detector monitors the absorption of RF energy, and the electronic signal is then amplified and displayed as a peak. Nowadays, a more popular method is the Fourier transform method that requires a radiofrequency pulse of a short time, which covers the whole frequency range for the nucleus being measured. The stimulated spin nuclei reach a new equilibrium state upon the irradiation of the radiofrequency and relax back to the original state to generate an oscillating magnetic field right after the pulse is removed. A receiving coil collects the decaying signals, called *free induction decays* (FIDs), which will eventually be mathematically Fourier transformed to the spectral signals to build up a NMR spectrum.

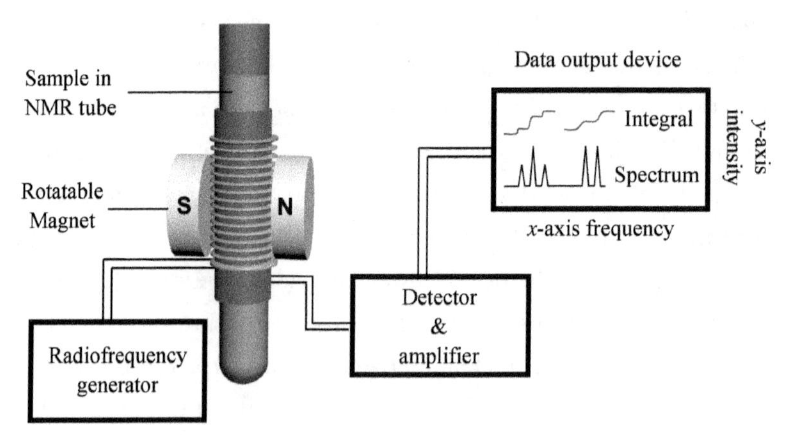

Figure 4.17 Schematic representation of a NMR spectrometer

To minimize the interference from the same type of nuclei of other compounds, purity of the tested compound should always be as high as possible. As liquid ^{1}H NMR is much more convenient to undertake and deuterated solvents are typically chosen to dissolve the sample in order to exclude the interference of ^{1}H atoms present in the solvent before the ^{1}H NMR experiment. To some point, the residual hydrogen signals in the deuterated solvent can also function as an internal reference of the spectrum in addition to tetramethylsilane and Table 4.6 summarizes the chemical shift information of several frequently used deuterated solvents.

In addition, since the homogeneity of the solution is important for better quality of the NMR data, clean and specific NMR tubes must be used and insoluble particles must be removed before the test.

Table 4.6 Chemical shifts of some frequently used deuterated solvents

Solvent formula	^1H NMR		^{13}C NMR
	δ (ppm)	Peak splitting	σ (ppm)
$CDCl_3$	7.24	triplet	77.0
CD_3COCD_3	2.04	septet	29.8
CD_3CN	1.93	septet	1.3, 117.7
CD_2Cl_2	5.32	quintet	53.5
CD_3SOCD_3	2.49	septet	39.7
CD_3OD	3.35, 4.78	septet	49.3
D_2O	4.65	—	—

2. *Experimental*

1. Make sure the sample from Experiment 5.20 is dry and pure[①]. Dissolve the sample into 0.3~0.7 mL of a suitable deuterated solvent.

2. Make sure insoluble solid impurities have been filtered out, and then transfer the solution into a clean NMR tube, and seal it with a NMR cap tightly. The sample tube in a 5 mm diameter must be of uniform outside and inside diameter and with uniform wall thickness. And it's a good habit to label your tubes clearly.

3. Place the NMR tube with the sample solution from Experiment 5.20 in a suitable deuterated solvent between the poles of a magnet, then carry out a ^1H NMR experiment.

3. *Question*

Predict the ^1H NMR spectrum of cyclohexene (product from Experiment 5.20) before the laboratory.

① Make sure to remove the residual solvents in the sample after recrystallization, wash, or flash column chromatography.

5 Organic Experiments

5.1 Purification of Liquids by Distillation

Objectives

1. To master the technique of simple distillation and apply distillation on isolation of liquid mixtures.

2. To determine the concentration of the aqueous EtOH solution provided.

Hazards

1. Chemical hazards

Aqueous ethanol.

2. Physical hazards

Heating might cause skin-burning and broken glassware may lead to skin damage.

Equipment

1. Boiling chips.
2. Distillation kits (distillation head, Liebig condenser, receiver adapter, thermometer with adapter).
3. 50 mL Erlenmeyer flasks that fit the receiver adapter.
4. 10 mL or 50 mL Graduated cylinder.
5. Glass wool and aluminium foil (optional).
6. Heating set-up.
7. Iron stand and clamps.
8. 25 mL & 50 mL Round bottomed flask.
9. Rubber tubing.

Pre-lab work

Before entering the wet lab, read thoroughly about the experiment and understand the physicochemical principle behind. Plan your set-up of equipment by visiting useful sources and complete your pre-lab report.

Introduction

It is widely known that there are generally three basic states for any chemical: Gas, liquid, and solid. The phase transition from one state to another occurs at a certain temperature under a certain atmosphere and involves heat transfer, either exothermically or endothermically. Such phase conversion has been widely used for compound separation. Distillation is one of the techniques making use of the phase transition between the liquid phase and the gas phase for separation of a liquid mixture. The efficiency of distillation highly depends on the difference in boiling points of the components present in the mixture. During distillation, evaporation of liquid and cooling of vapor are associated to achieve separation. The process of preparing traditional liquor is a classic example of distillation. Another common usage of distillation is in the purification of crude oil for gasoline. Noticeably, both processes involve the purification of mixtures of LIQUIDS. As in the organic field, distillation is the method primarily used to purify compounds from a liqud mixture.

How does distillation work to purify liquids? The theoretical ideas about the process are related to what has been covered in General Chemistry and Physical Chemistry with regard to vapor pressure equilibrium. In order to employ the technique, it is important for one to understand how vapor pressures of mixtures depend on the structures of the components. Additionally, it is useful for one to understand how vapor pressure of each component affects the distillation and how it facilitates the separation.

Try to imagine there is a flask in front of you at room temperature containing a certain amount of a liquid. It is known that the molecules in the liquid are dynamic. Namely, they are in constant motion and have a certain degree of kinetic energy. Some of these molecules may own enough internal kinetic energy so that they can actually escape from the liquid phase to the gas phase. At any temperature, this takes place to a certain degree. Over time a certain amount of the liquid molecules will escape into the gas phase and exert a pressure (vapor pressure) on the liquid below. Although vapor pressure differs for different temperatures, vapor pressure always increases as the temperature rises.

As the temperature of the system increases (say, by externally heating the flask), more liquid molecules will absorb enough energy to escape to the gas phase. Accordingly, the vapor pressure will also increase. Ultimately, we expect that as we raise the temperature, the vapor pressure of the system increases to the point where it equals the external pressure, the atmospheric pressure at the liquid surface in addition to hydrostatic pressure on the system. At this temperature, always referred as the boiling point, the thermal energy has been enough for molecules in the interior of the liquid to enter the gas phase and form bubbles that can rise to the surface of the liquid. The result is the observed phenomenon of boiling. As different chemicals have distinct boiling points, controlling the heating power can allow one or several substances over the others to turn from the liquid phase to the gas phase for separation. The process of distillation produces vapor by boiling a liquid and subsequently directs the vapor into a separate apparatus for condensation and collection. The apparatus for this process can vary considerably depending upon the type of separation that one desires. There are basically two types of distillation set-up—simple and fractional. In this experiment, the simple distillation① is involved.

A note on azeotropes

Not all liquids conform to the *Raoult's law* and form ideal solutions. Some chemicals can co-evaporate to afford a vapor with a constant composition at a unique and constant boiling point, which is different from and always lower than the boiling point of any of the chemicals present in the mixture. The vapor is condensed to afford a liquid mixture, which is called azeotrope. One needs to be careful when dealing with azeotropic systems, because a constant boiling point is not necessarily an indication of a single component system. In fact, an azeotrope represents an example of a system that consists of two (or even three) components that boil at a constant temperature, which is different from the boiling point of either pure component.

The distillation associated with azeotropes is often referred to azeotropic distillation. An example is the mixture of water and ethanol. Because of the intermolecular interactions between water and ethanol molecules, a unique vapor mixture (an azeotrope) containing 95.5% ethanol and 4.5% water forms while aqueous ethanol boils at 78.1 ℃, which is lower than that of ethanol (78.4 ℃). No matter how efficient the distilling apparatus is or how hard one is trying, pure 100% ethanol cannot be obtained by the simple distillation of a mixture of ethanol/water.

Procedure for Distillation

Weigh ~33 mL aqueous ethanol of an unknown concentration in a 50 mL round bottom flask containing several boiling chips. Before heating the flask, assemble the glassware and start gently the circulation of the cooling water in the condenser. When everything is ready, adjust the heat powered by a Variac, so that INITIALLY the liquid boils rapidly. Then maintain a gentle boiling of the mixture in the flask, so be

① Depending on the external pressure applied to the distillation system, the simple distillation can be undertaken at one atmospheric pressure or lower. The former is referred as distillation at atmospheric pressure and the latter is vacuum distillation. The former kind is practice in this experiment.

advised that one may have to adjust the setting on the Variac. *Do not heat up too quickly otherwise a good separation of components will not be achieved.* You are to collect the distillate in a pre-weighed collection flask. ONE NEED TO DETERMINE WHEN TO COLLECT and must record the temperature at the moment when collecting the aliquots. NEVER distill the flask to dryness. A rapid temperature drop from the thermometer readout is a good sign of the endpoint for one component. Once the distillation is complete, turn off the heat immediately, weigh the collection flask with distillate, record the appearance of the distillate, and submit the distillate to the instructor.

During the experiment, one should keep a good and accurate record of the experimental procedure and observations in the notebook. Clean-up and check-out before leaving the lab.

Technique Checklist

1. Setting up distillation glassware correctly

(1) All joints should be attached tightly. Clicks might be used to guarantee this.

(2) The bottom bulb of the thermometer should be located at the branch of the distillation head, which measures the temperature of the vapor that is condensed and then collected in the collection flask.

2. Performing atmospheric pressure distillations

(1) The running water should be turned on before the distillation starts and the water flows in at the bottom inlet of the condenser and out at the top outlet of the condenser.

(2) A higher heat power is applied at the beginning but a gentle heat power is used later on for distillation to avoid overheating the flask.

(3) During distillation, the thermometer readout should be closely monitored and the number should be recorded when distillate starts to drop out.

(4) Never heat the distillation flask to dryness.

Pre-lab Discussion

1. Theory of distillation.
2. Distillation glassware and how to set it up.

Experiment Outline

1. Use a graduated cylinder to measure aqueous ethanol and weigh it by a taring method or a subtraction method.
2. Assemble the glassware for distillation.
3. Perform atmospheric pressure distillation.
4. Record the mass of the purified low-boiling compound and its boiling point.

Helpful Hints

1. Ensure all joints are sealed well. Otherwise, the product may fly into the atmosphere.

2. Do not heat the mixture too fast, or the entire sample may end up in the collection flask.

3. Insulate the distillation head with cotton and foil to facilitate the distillation.

4. Be aware that the temperature reading on the thermometer may not correlate accurately with the boiling point of the distilling liquid. The temperature at steady state usually varies in a small range.

Results

1. Record the amount of the aqueous ethanol used in the experiment.

2. Record the boiling temperature range.

3. Record the amount of the purified liquid.

4. Record the appearance of the purified liquid.

5. Determine the weight percentage of the aqueous ethanol provided.

Hint: Assume that the overall mass of ethanol remains the same in the distillate as in the initial aqueous sample used for the distillation, and an azeotrope is collected during the distillation.

5.2 Purification of Organic Compounds by Extraction

Objectives

1. To master extraction and apply extraction on isolation of acetic acid from the aqueous solution.

2. To determine the recovery ratio of acetic acid by titration.

3. To determine the partition coefficient K_p of acetic acid between diethyl ether[①] and water.

4. To compare the extraction efficiency between a single extraction and multiple extractions using the same total volume of solvent.

Hazards

1. Chemical hazards

Aqueous acetic acid, diethyl ether, 0. 2000 mol • L^{-1} standardized NaOH

① Ethyl acetate or some other water-immiscible solvent can be used instead of diethyl ether.

solution, 0.1% phenolphthalein in 95% ethanol (indicator).

2. Physical hazards

Broken glassware may lead to skin damage. NO OPEN FLAME is allowed while diethyl ether is used.

Equipment

1. 50 mL Burette, Burette clamp, and support stand.
2. 50 mL Erlenmeyer flasks.
3. 50 mL Graduated cylinder.
4. Iron ring, iron stand and clamps.
5. 50 mL or 125 mL Separatory funnel (sep funnel).
6. 5 mL and 10 mL Transfer pipettes.

Pre-lab work

Before entering the wet lab, read thoroughly about the experiment and understand the physicochemical principle behind. Plan your set-up of equipment by visiting useful sources and complete your pre-lab report.

Introduction

Extraction is one of the most used purification techniques in organic labs. It is extremely useful for some thermo-sensitive compounds while distillation at high temperature may cause such compounds to decompose. In addition, recrystallization is sometimes powerless to some co-crystallization systems and extraction may help to get rid of one over the other by manipulating the acidity of the aqueous phase. Extraction is based on compound partitioning in two immiscible phases and obeys the "*like likes like*" principle. Based on the partitioning equation, it is easy to deduce that extraction by several times of a solvent is usually more efficient than the extraction by the same total amount of the solvent once.

Procedure

Extraction 1 and Titration 1: 10 mL of aqueous acetic acid is transferred with a transfer pipette to a sep funnel with its stopcock closed. 30 mL of diethyl ether is transferred to the sep funnel. Then stopper the funnel, invert it, and release its pressure by opening the stopcock. Close the stopcock, invert and shake the sep funnel, and vent until no audible or visible gas emerges (*i. e.* nothing comes out of the stopcock when releasing pressure). **Place the sep funnel in an iron ring on a ring**

stand and remove the stopper immediately. [①] Allow the layers to separate thoroughly to give a clear and stable interface. Then drain the lower layer of solution (Is that aqueous or organic?) into a labeled Erlenmeyer flask. Be careful and slow to drain as the liquid-liquid interface lowers in the funnel. Be sure to leave a drop of the bottom layer in the sep funnel. Add 3～4 drops of 0. 1% phenolphthalein solution to the Erlenmeyer flask and titrate the aqueous solution with 0. 2000 mol • L^{-1} NaOH.

Extraction 2 and Titration 2: 10 mL of the aqueous acetic acid is transferred with a transfer pipette to a clean sep funnel with its stopcock closed. Then 10 mL of diethyl ether is transferred to the sep funnel. Undertake extraction as described above and then separate the layers. The aqueous layer is further extracted with 10 mL of fresh diethyl ether twice in the sep funnel. The aqueous layer is collected and titrated with 0. 2000 mol • L^{-1} NaOH using 3～4 drops of 0. 1% phenolphthalein solution as the indicator.

Titration 3: 5 mL of the aqueous acetic acid is transferred with a transfer pipette to an Erlenmeyer flask and titrated with 0. 2000 mol • L^{-1} NaOH using 3～4 drops of 0. 1% phenolphthalein solution as the indicator.

At the end of the experiment, all organic layers are poured into a waste solvent container while the aqueous solutions are poured into the drain. Clean-up and check-out before leaving the laboratory.

Technique Checklist

1. Liquid transfer.
2. Extraction.
3. Titration.

Pre-lab Discussion

1. Theory of extraction.
2. Sep funnel and how to work with it.
3. Titration.

Experiment Outline

1. Two series of extraction are conducted.

(1) For either extraction, 10 mL of acetic acid is transferred to the sep funnel.

(2) For one of the extraction, 30 mL of diethyl ether is added to the sep funnel.

(3) For the other extraction, 10 mL of diethyl ether is added to the sep funnel. When the aqueous layer and the diethyl ether layer are separated, the extraction is

① One can also wait to remove the stopper until draining the lower layer.

repeated with 10 mL of fresh diethyl ether twice.

2. Titrate the aqueous layers with NaOH standard solution.

Helpful Hints

1. Absolutely **NO** open flame is allowed.

2. Hold the sep funnel firmly with your palm against the stopper.

3. Relieve the internal pressure each time after shaking the sep funnel with solutions.

4. DO NOT POINT THE MOUTH OF SEP-FUNNEL TO ANYONE.

5. Wait till the interface of the two phases becomes clear to separate.

6. Be careful and slow to drain the lower layer as the interface lowers in the funnel.

Results

1. Record the volumes of NaOH solution used for the titration of the aqueous layers from the extraction and aqueous acetic acid without titration.

2. Calculate the recovery ratio of either strategy.

3. Compare the extraction efficiency of the two extraction strategies.

4. Determine the partition coefficient K_p for acetic acid between diethyl ether and water at the experimental temperature.

5.3 Purification of Solids by Recrystallization

Objectives

1. To master recrystallization to purify solid organic compounds.

2. To practice the dissociation of impurity from the sample with minimum amount of an appropriate solvent at high temperatures.

3. To master vacuum (suction) filtration.

4. To identify the purity of product by thin layer chromatography (TLC).

Hazards

1. Chemical hazards

N-Phenyl acetamide, activated carbon, water.

2. Physical hazards

Heating might cause skin-burning and broken glassware may lead to skin damage.

Equipment

1. Boiling chips.
2. Büchner funnel and filter flask with a rubber adapter.
3. Condenser.
4. 250 mL Erlenmeyer flask with ground glass joint.
5. Filter paper.
6. 100 mL graduated cylinder.
7. Heating set-up.
8. Iron stand and clamps.
9. Rubber tubing.
10. Stir rod or spatula.
11. TLC plate① and capillary tube.
12. TLC chamber and aluminium foil.
13. UV light at 254 nm.
14. Water aspirator or water pump.
15. Weighing paper.

Pre-lab work

Before entering the wet lab, read thoroughly about the experiment and understand the physicochemical principle behind. Plan your set-up of equipment by visiting useful sources and complete your pre-lab report.

Introduction

Recrystallization

Recrystallization is an important method to purify solid compounds, which have drastic solubility change in a solvent along temperature variation. This technique usually involves the preparation of a saturated solution at high temperature, such as the boiling point of the solvent used, and crystallization upon cooling the clear solution. Through this technique, various solid compounds can be further purified in scales of grams or higher.

The recovery yield is always used to evaluate the purity of the sample by recrystallization:

$$\%recovery\ of\ solid = \frac{m(solid) - m(solid\ lost)}{m(solid)} \times 100$$

① The TLC plate recommended in this book is coated with silica gel and fluorescing F_{254}.

One should be aware of that more solvent used than necessary always results in a low recovery rate. Therefore, it is better to minimize the amount of solvent during recrystallization. After the dissolution at higher temperature, the resulting solution should be allowed to cool by itself without interruption. A rapid cooling by immersing the flask containing the solution in a cold water bath may result in the capsulation of impurities in the compound crystals and tiny crystals may form because of the sudden and rapid crystallization. If no crystal appears upon cooling, addition of some crystals to the cooled oversaturated solution can stimulate crystallization with nucleation cores present. Scratching the inner wall of the flask with a stir rod or spatula can also help to provide some nucleation cores for the same purpose.

Procedure

1. *Recrystallization*

Weigh ~5 g of impure *N*-phenyl acetamide into a 250 mL Erlenmeyer flask, to which a desired amount of water (~half of the water that is required for 5 g of pure compound to dissolve at the high temperature) was then added. The mixture in the flask is heated to boil while being stirred①.

Remove the flask from the heating source and examine the solution. If there are particles present, continue to heat the mixture to boil and a small amount of hot or cold water is added. Keep adding water in small amounts (several drops at a time from a Pasteur pipette) until all of the *N*-phenyl acetamide is dissolved in the boiling solution. Do not add too much water or the solution will not be saturated anymore and the recovery yield of purified *N*-phenyl acetamide will be reduced.

If there are particles in the solution obviously not *N*-phenyl acetamide, like dirt, hot filtration should be carried out to remove the particles.

If the hot solution is in dark color, activated carbon may be added carefully to decolor the solution. However, attention should be given that ADDING POROUS SOLID like activated carbon to a BOILING SOLUTION would result in violent boiling/bumping/splattering and lead to loss of materials.

The solution is then put aside and allowed to cool down to room temperature for slow crystallization. When the crystals have formed completely (may require an ice bath), collect the solid by setting up a vacuum (suction) filtration on a properly cut filter paper in a clean Büchner funnel. Pour the chilled mixture into the Büchner funnel. The water should pass through quickly. If not, check for vacuum leaks. Get

---

① Try to avoid the solid from settling down at the bottom of the container by stirring. If the solid melts into oily liquid, a vigorous stir should be applied to maximize the interaction area between the oil and water while additional water is added until all the oil disappears.

all the crystals out of the flask using a spatula or stirring rod. Rinsing the flask with 1 or 2 mL of **cold** water helps get the crystals out of the flask, and rinsing helps remove the residual solution on the surface of the crystals as well.

Let the aspirator run for a few minutes to air-dry the crystals. Then use a spatula to lift the filter paper and crystals out of the Büchner funnel, then press them as dry as possible on a large clean paper towel (hand dry), allow them to dry completely, and transfer the dry sample to a pre-weighed weighing paper. Record the weight and appearance of the DRY recovered *N*-phenyl acetamide crystals.

2. *TLC analysis*

Transfer several crystals into a clean and dry test tube and dissolve the crystals with minimum ethanol. Draw a line (the origin) ~0.5—1 cm to the lower edge of a TLC plate using a pencil. Then place one end of the capillary tube in the solution and the solution will rise up in the capillary. Gently touch the tip of the capillary tube with the place on the line of the TLC plate (usually in the middle) and allow the spot to dry prior to the TLC development. On the other hand, pour the TLC developing solvent in a beaker (or a TLC chamber) with a filter paper and make sure the solvent level is lower than the TLC origin, and soak the filter paper with the solvent. Cover the beaker with a piece of aluminium foil and let it sit still for 2—3 min. Put the TLC plate in the beaker and then cover the beaker with aluminium foil. Let the TLC plate develop in the chamber until the solvent (eluent) travels up the plate to ~0.5—1 cm away from the top. Take the plate out and mark the solvent front as quickly as possible. Let the plate dry and visualize the plate under UV light at 254 nm.

Two solvent systems are used for TLC analysis: 50% EtOAc/petroleum ether and 10% MeOH/dichloromethane, respectively.

Submit the crystals and TLC plates to the instructor. Clean-up and check-out before leaving the laboratory.

Technique Checklist

1. Recrystallization.
2. Vacuum filtration.
3. TLC analysis.

Pre-lab Discussion

1. Theory of recrystallization and TLC analysis.
2. Recrystallization set-up.
3. Vacuum filtration and set-up.
4. TLC analysis.

Experiment Outline

1. Weigh ~5 g of crude *N*-phenyl acetamide into an Erlenmeyer flask.

2. Perform recrystallization.

(1) Add some water to the solid.

(2) Heat the mixture of solid and water to boil.

(3) Slowly add more water in portions until all solid dissolves.

(4) Allow the solution to cool down on its own.

3. Collect the crystals by vacuum filtration.

(1) Rinse the flask with a small amount of water.

(2) Ensure the filter paper is cut to the size covering all holes in the Büchner funnel but not cover the side wall.

4. Dry and weigh the crystals.

5. TLC analysis.

Helpful Hints

1. It is known that ~0. 53 g of *N*-phenyl acetamide dissolves in 100 mL of water at room temperature (~25 ℃) while ~5. 5 g of the compound can dissolve in 100 mL of boiling water.

2. Make sure to heat the flask to temperatures lower than the melting point of *N*-phenyl acetamide (111—113 ℃).

3. Do not add boiling chips or activated carbon to the boiling solution abruptly.

4. Scrape the sides of the flask to stimulate the formation of crystals if necessary.

5. Use appropriate sized filter paper for vacuum filtration.

6. Do not overdevelop the TLC plate.

7. Measure the distance between the orign and the solvent front (position the sample runs) after the TLC development.

Results

1. Record the weight of crystals obtained.

2. Describe the crystal appearance: Shape and color.

3. Determine the purity of the obtained crystals by the TLC analysis.

4. Calculate R_f values and the recovery yield of the purified crystals.

5.4 Purification of *Cyclo*hexane

Objectives

1. To practice extraction and distillation in compound purification.
2. To check the presence of phenol by a ferric test.

Hazards

1. Chemical hazards

Phenol contaminated *cyclo*hexane, 1% $FeCl_3$ aqueous solution, 5% NaOH solution, water, anhydrous $CaCl_2$.

2. Physical hazards

Heating might cause skin-burning and broken glassware may lead to skin damage.

Equipment

1. Distillation kit (distillation head, Liebig condenser, receiver adapter, thermometer with adapter).
2. 50 mL Erlenmeyer flasks.
3. 25 mL Graduated cylinder.
4. Heating set-up.
5. Iron ring, iron stand and clamps.
6. Liquid funnel and filter paper.
7. 50 mL Round bottom flask.
8. Rubber tubing.
9. 50 mL or 125 mL Separatory funnel.
10. 5 mL Test tubes and Pasteur pipettes.

Pre-lab work

Before entering the wet lab, read thoroughly about the experiment and understand the physicochemical principle behind. Plan your set-up of equipment by visiting useful sources and complete your pre-lab report.

Introduction

*Cyclo*hexane is one of the most used organic solvents and roughly 90% of *cyclo*hexane is used in manufacture of nylon or related industry. It can be produced by catalytic hydrogenation of benzene in either liquid phase or gas phase. *Cyclo*hexane can also be produced from phenol via a two-step process. However, phenol contamination is a common problem in the second industrial production pathway. In order to lower the cost of separation and recycle phenol for the reaction, washing *cyclo*hexane with basic aqueous solutions is applied to get phenol into the aqueous phase, which can later be released by an acid treatment. The *cyclo*hexane portion, which contains no phenol and is dried, can then be further isolated using simple distillation.

Procedure

The presence of phenol should be first identified by adding 1 drop of 1% $FeCl_3$ aqueous solution to 0.2 mL of the provided *cyclo*hexane and 0.2 mL of water in a 5 mL test tube. Weigh 20 mL of *cyclo*hexane and transfer it to a separatory funnel. Add 20 mL of 5% NaOH solution to the separatory funnel and the two layers are thoroughly mixed by shaking the separatory funnel several times and the separatory funnel is placed in a ring stand to allow the layers to separate. The aqueous layer is drained and discarded. A second portion of 20 mL of 5% NaOH solution is added to the separatory funnel to wash the organic layer again. The organic layer is subsequently washed with 20 mL of water and dried with anhydrous $CaCl_2$. The solid is filtered off and the filtrate is subject to simple distillation to collect the fraction of 80—81 ℃ in a pre-weighed collection flask. The mass of the collection flask containing distillate is then measured after the distillation. In a clean 5 mL test tube, 1 drop of 1% $FeCl_3$ solution is added to 0.2 mL of the distillate and 0.2 mL of water to check if any residual phenol is present in the distillate.

Submit the distillate to the instructor. Clean-up and check-out before leaving the laboratory.

Technique Checklist

1. Undertaking the ferric test.
2. Performing extraction and distillation.
3. Performing gravity filtration.

Pre-lab Discussion

1. Extraction and distillation.
2. Principle for the ferric test.

Experiment Outline

1. Wash 20 mL *cyclo*hexane sample with 20 mL 5% NaOH solution twice and then 20 mL water.

2. Dry the organic layer with anhydrous $CaCl_2$.

3. Perform distillation.

Helpful Hints

1. Phenols and enols can be identified by ferric treatment and the enols can form unique colorful complexes with ferric.

2. Phenol is acidic and reacts with NaOH to give water soluble sodium phenolate.

3. If emulsion forms during extraction, sodium chloride can be added to break the emulsion.

4. Boiling point of *cyclo*hexane is 80.7 ℃.

Results

1. Record the weight of organic liquid before and after the extraction and/or distillation.

2. Record the boiling point of the organic compound.

3. Record the phenomenon for the ferric test before and after the extraction and/or distillation.

4. Record the appearance of the final product.

5. Calculate the recovery yield.

5.5 Preparation of Bromoethane

Objectives

1. To prepare alkyl halide by an S_N2 reaction.

2. To understand the mechanism of the S_N2 reaction.

3. To practice simple distillation and extraction.

Hazards

1. Chemical hazards

NaBr, concentrated H_2SO_4, 95% EtOH①, H_2O.

2. Physical hazards

Broken glassware may lead to skin damage and heating might cause skin-burning.

Equipment

1. Boiling chips.

2. Distillation kit (distillation head, thermometer and adapter, Liebig condenser,

① This experiment can use butan-1-ol to replace ethanol for 1-bromobutane in the same way.

and receiver adapter).

3. 100 mL or 50 mL Erlenmeyer flasks.
4. 10 mL or 25 mL Graduated cylinder.
5. Heating set-up.
6. Ice-water bath.
7. Iron ring, iron stand and clamps.
8. Pasteur pipettes.
9. 100 mL Round bottom flask.
10. Rubber tubing.
11. 50 mL or 125 mL Separatory funnel (sep-funnel).
12. 10 mL Transfer pipettes.

Pre-lab work

Before entering the wet lab, read thoroughly about the experiment and understand the physicochemical principle behind. Plan your set-up of equipment by visiting useful sources and complete your pre-lab report.

Introduction

Alkyl halides are useful reagents in organic synthesis. There are several ways to prepare alkyl halides, and preparation from corresponding alcohols is one of the most used methods. Take the preparation of ethyl bromide for instance, in a synthetic laboratory it is usually synthesized by treating concentrated hydrobromic acid with ethanol. Hydrobromic acid is highly corrosive and may cause various health issues when exposed. On the other hand, hydrobromic acid can be generated *in situ* by mixing sodium bromide with concentrated sulfuric acid. Accordingly, the preparation of ethyl bromide can be achieved by treating ethanol with sulfuric acid and sodium bromide. In fact, the reaction between ethanol and hydrobromic acid is reversible. To push the reaction forward, we can a). increase the concentration of one of the starting materials, and b). remove the product quickly from the reaction mixture once it has been formed. Both procedures would facilitate the forward reaction and lead to an increasing conversion yield.

$$NaBr + H_2SO_4 \longrightarrow HBr + NaHSO_4$$

$$CH_3CH_2OH + HBr \rightleftharpoons CH_3CH_2Br + H_2O$$

However, there are several side-reactions accompanying to provide diethyl ether, ethylene, and even bromine[①]. To avoid the side-reactions between ethanol and

① For 1-bromobutane synthesis, dibutyl ether would be the major side product.

sulfuric acid, diluted sulfuric acid is used to decrease the possibility of ether and ethylene production. To avoid the oxidation of hydrobromic acid by sulfuric acid, diluted sulfuric acid and also gentle heating is adopted. Bromoethane has a boiling point of 38.4 ℃ under 1 atmosphere① and can be collected by simple distillation during the reaction to allow the removal of the product and facilitate the forward reaction.

Procedure

To a 100 mL round bottom flask is added 10 mL (0.17 mmol) of 95% ethanol and 9 mL of water. The mixture is cooled with an ice-water bath and 19 mL (0.34 mol) of concentrated sulfuric acid is slowly added while shaking. After the completion of the addition, the obtained mixture is allowed to cool to room temperature and grounded sodium bromide (15 g) is added to the solution when stirring.② Several boiling chips are then added, and the round bottom flask is assembled with the distillation head, thermometer (and adapter), Liebig condenser, the receiver adapter,③ and a collecting Erlenmeyer flask that contains ~10 g crushed ice and is immersed in an ice-water bath.④ The round bottom flask is then heated in a water bath of ~60—70 ℃.⑤ The process should be terminated when there is no more product dropping into the collection flask, which may take ~40 min.

The distillate is transferred to the sep funnel and the organic layer is collected in a dry Erlenmeyer flask in an ice-water bath. Then, 2 mL of concentrated sulfuric acid is carefully dropwise added to the product while the flask is gently shaken. Then the mixture is stored still to allow separation of layers and transferred to a **dry** sep-funnel to separate layers. The organic layer is transferred to a dry round bottom flask with several boiling chips and heated under water bath for simple distillation.⑥ The product is collected in a dry and pre-weighed Erlenmeyer flask, which is immersed in an ice-water bath. Weigh the Erlenmeyer flask containing distillate after water at the external surface of the flask is wiped off.

Submit the distillate to the instructor. Clean-up and check-out before leaving the laboratory.

---

① The boiling point of 1-bromobutane is 101.4 ℃.

② For the synthesis of 1-bromobutane, it is suggested to use 20 mL of water, 20 mL of sulfuric acid, 13 mL of butan-1-ol, and 17 g of sodium bromide.

③ A rubber tube connected with a dry tube filled with sodium carbonate is recommended to attach to the receiver adapter to prevent HBr from flying to the atmosphere.

④ A regular refluxing set-up should be used for the synthesis of 1-bromobutane and an absorption set-up should attached to the refluxing condenser for the exhausted gas.

⑤ The reaction for 1-bromobutane is heated to reflux for 45 min instead.

⑥ An oil bath should be used for the distillation of 1-bromobutane.

Technique Checklist

1. Using transfer pipette to measure a desired amount of a liquid reagent.
2. Collecting crude in an ice-water bath.
3. Performing extraction with sep funnel.
4. Drying the organic layer with concentrated sulfuric acid.
5. Performing distillation while the collection flask is cooled in an ice-water bath.

Pre-lab Discussion

1. Mechanism of the synthesis of bromoethane from ethanol.
2. Possible side-products and reasons.
3. How to obtain the pure product.

Experiment Outline

1. Measure ethanol (10 mL) to a round bottom flask by a transfer pipet.
2. Measure sulfuric acid and water by a graduated cylinder.
3. Slowly add sulfuric acid to the aqueous ethanol solution using a Pasteur pipette.
4. Cool the mixture to room temperature before the addition of NaBr.
5. Perform atmospheric pressure distillation and extraction.
6. Obtain the mass of the crude product and purified product.
7. Record the boiling point of the product.

Helpful Hints

1. Make sure all joints of glassware are sealed well. Otherwise, the product would fly into the atmosphere.

2. Do not overheat the mixture to avoid a poor conversion yield and severe side reactions.

3. Use an ice-water bath to cool the fraction in the collection flask to avoid further loss of the product by evaporation.

Results

1. Record the boiling point and amount of the crude and purified liquid.
2. Record the phenomena accompany your procedures.
3. Record the appearance of the crude and final product.
4. Calculate the isolated yield of the reaction.

5.6 Synthesis of Ethyl Acetate

Objectives

1. To synthesize ethyl acetate by the Fischer esterification.
2. To practice extraction, filtration, and distillation.

Hazards

1. Chemical hazards

Ethanol, acetic acid, Conc. H_2SO_4, saturated sodium bicarbonate solution, saturated calcium chloride solution, anhydrous magnesium sulfate, brine.

2. Physical hazards

Heating might cause skin-burning and broken glassware may lead to skin damage.

Equipment

1. Adapters.
2. Boiling chips.
3. Condenser.
4. Distillation kit (distillation head, thermometer and adapter, Liebig condenser, and receiver adapter).
5. 10 mL Graduated cylinder.
6. Heating set-up.
7. Iron ring, iron stand and clamps.
8. Liquid funnel and filter paper.
9. Litmus paper.
10. 50 mL Round bottom flasks.
11. Rubber tubing.
12. 125 mL or 250 mL Separatory funnel.

Pre-lab work

Before entering the wet lab, read thoroughly about the experiment and understand the physicochemical principle behind. Plan your set-up of equipment by visiting useful sources and complete your pre-lab report.

Introduction

Esterification is a type of organic reactions to make esters. Preparation of esters from alcohols and carboxylic acids usually uses an acid as a catalyst and this reaction is known as the Fischer esterification. The preparation of ethyl acetate is a classic example of the Fischer esterification, and the reaction equation is shown below. Such reaction is reversible and quickly reaches an equilibrium. In order to push the reaction forward, the by-product water is usually removed from the system by dehydrating reagent or azeotropic distillation with some solvent or reagent. For instance, conc. H_2SO_4 is usually used as the catalyst but its dehydrating capability can help to drive the reaction forward.

$$CH_3COOH + CH_3CH_2OH \xrightleftharpoons[\text{refluxing}]{H_2SO_4} CH_3COOCH_2CH_3 + H_2O$$

Procedure

In a 50 mL round bottom flask are mixed 9. 5 mL (~0. 2 mol) of anhydrous ethanol and 6 mL (~0. 10 mol) of acetic acid. Then 2. 5 mL of concentrated sulfuric acid is slowly and carefully added. After being mixed well, the mixture is added with boiling chips and the flask is attached with a condenser. The reaction is slowly heated to reflux for 0. 5 h and subsequently cooled to an ambient temperature. The mixture is then subject to simple distillation and the distillate is collected with a flask immersed in a cold water bath. The distillation is terminated when half of the reaction volume (~9 mL) is collected.

Saturated sodium bicarbonate solution is slowly added to the distillate with constantly shaking until no further gas is released or the litmus paper suggests the aqueous solution be no longer acidic. The mixture is transferred to a separatory funnel and the aqueous layer is drained. The organic layer is washed with 5 mL of brine, 5 mL of saturated calcium chloride, and 5 mL of water successively. Then the organic layer is collected in a pre-weighed Erlenmeyer flask. The Erlenmeyer flask is weighed for the crude weight and dried with minimum anhydrous magnesium sulfate for 10—15 min. The solid is removed by filtration, and the organic layer is subject to distillation for the fraction at 73—78 ℃ in a pre-weighed dry Erlenmeyer flask. Weigh the Erlenmeyer flask again to obtain the product mass.

Submit the distillate to the instructor. Clean-up and check-out before leaving the laboratory.

Technique Checklist

1. Setting up a reaction by refluxing.
2. Assembling distillation glassware correctly.
3. Performing extraction and simple distillation.
4. Drying an organic solution with anhydrous magnesium sulfate.

Pre-lab Discussion

1. Mechanism of the Fischer esterification.

2. The purpose of using saturated sodium bicarbonate solution and saturate calcium chloride solution.

Experiment Outline

1. Use graduated cylinders to measure ethanol (9. 5 mL), acetic acid (6. 5 mL), and sulfuric acid (2. 5 mL) to a dry round bottom flask.

2. Heat the reaction to reflux for 0. 5 h.

3. Perform simple distillation and extraction.

4. Obtain the mass of the crude product and purified product.

Helpful Hints

1. Make sure all your joints are sealed well during refluxing or distillation. Otherwise, the product will fly into the atmosphere.

2. Do not heat the mixture subject to distillation too fast, or the entire sample may end up in your collection flask.

Results

1. Record the boiling temperature and amount of the crude and purified liquid.

2. Record the phenomenon during the experiment.

3. Record the appearance of the product.

4. Calculate the isolated yield of the reaction.

5.7 Synthesis of Benzoin

Objectives

1. To understand the mechanism of the benzoin condensation and the role of thiamine in the reaction.

2. To practice recrystallization and TLC analysis.

3. To practice melting point analysis.

Hazards

1. Chemical hazards

Benzaldehyde, thiamine (vitamin B_1), ethanol, water, activated carbon, sodium hydroxide, 20% ethyl acetate in petroleum ether

2. Physical hazards

Heating might cause skin-burning and broken glassware may lead to skin damage.

Equipment

1. Boiling chips.
2. Büchner funnel and filtration flask.
3. Condenser.
4. Filter paper and rubber adapter.
5. 10 mL, 25 mL or 50 mL Graduated cylinder.
6. Heating set-up.
7. Ice-water bath.
8. Iron stand and clamps.
9. Melting point apparatus and capillary tubes for melting point test.
10. Pasteur pipette.
11. 50 mL Round bottom flask.
12. Rubber tubing.
13. 10 mL Test tubes.
14. TLC plates and capillary tubes.
15. TLC chamber and aluminium foil.
16. UV light at 254 nm.
17. Water pump.

Pre-lab work

Before entering the wet lab, read thoroughly about the experiment and understand the physicochemical principle behind. Plan your set-up of equipment by visiting useful sources and complete your pre-lab report.

Introduction

Benzoin, one of the well-known α-hydroxyl ketones, is an organic compound with a structure formula shown beside. The compound usually appears off-white crystals with a light camphor-like odor. Benzoin is a key precursor to benzil, which performs as a commonly used photo-initiator for polymerization. Benzoin can be synthesized from benzaldehyde under basic conditions via condensation in the presence of a catalyst, which is called the **benzoin condensation.** The benzoin condensation is a classic reaction between aldehydes containing no α-H to prepare α-hydroxyl ketones.

OH

O

Benzoin

Benzoin was first prepared by Justus von Liebig and Friedrich Wöehler during their study on the oil of bitter almond. It was found that traces of hydrocyanic acid facilitates the condensation of benzaldehyde. Such catalytic synthesis was later improved by Nikolay Zinin in the same era.

The benzoin condensation usually requires a nucleophile, such as cyanide anion, to catalyze the reaction. However, cyanide raises critical safety and environment issues due to its extreme toxicity under acidic conditions. It has been reported that only 1.5 mg of cyanide per kilogram of body weight can lead an adult to death just 1 minute after administration. Hence, organic scientists have thrived to find replacement of cyanide. Vitamin B_1, also known as thiamine or thiamin, has been found to behave similarly as cyanide in the benzoin condensation. The thiazole ring is positively charged, which demonstrates strong electron-withdrawing effect and hence makes the imine hydrogen in the thiazole ring very acidic. Upon deprotonation by a base as strong as sodium hydroxide, a carbon anion is formed. The carbanion hence can play the role as cyanide does in a traditional benzoin condensation to initiate the reaction. The mechanism of the benzoin condensation catalyzed by vitamin B_1 is as follows:

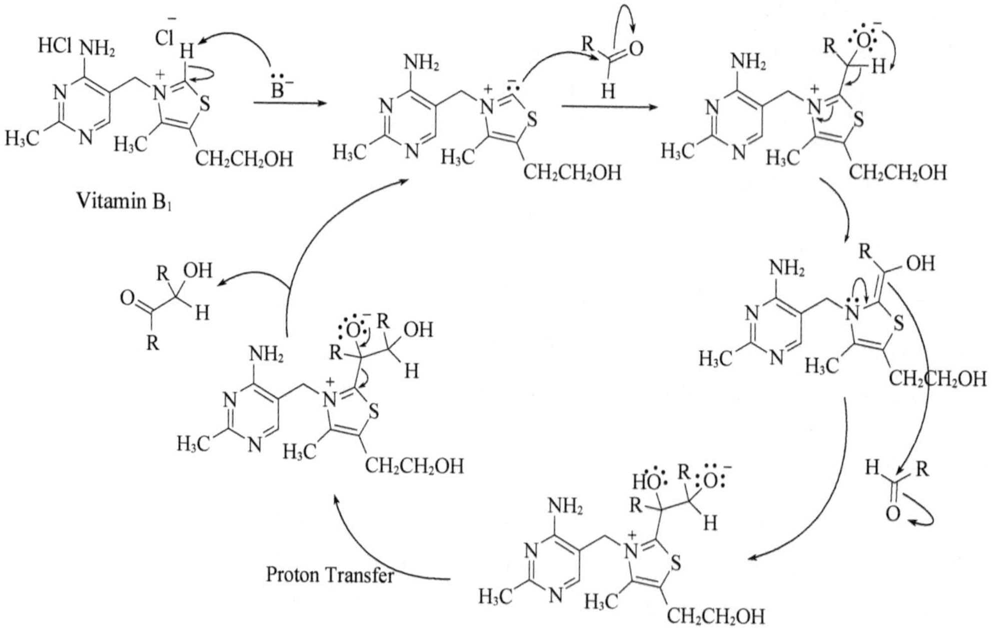

Vitamin B_1

Proton Transfer

Procedure

To a 50 mL round bottom flask are added 1.75 g of thiamine hydrochloride (vitamin B_1), 3.5 mL of water, and 15 mL of 95% ethanol. After turning homogeneous, the mixture is cooled over an ice-water bath. Meanwhile, 5 mL of 10%

sodium hydroxide solution in a test tube is cooled the same way. The cold sodium hydroxide solution is dropwise added to the mixture over ~2 min while the flask is maintained in the cooling bath. Upon completion of the sodium hydroxide solution addition, 10 mL (~10. 5 g) of fresh benzaldehyde is added. The resulting mixture is shaken gently to be thoroughly mixed and the pH of the solution is adjusted to 9—10. Subsequently, the ice-water bath is removed and the mixture is incubated in a water bath at 60—75 ℃ for 1 h right after a few boiling chips are added to the mixture. Then the reaction is heated to 90 ℃ for 0. 5 h. (The reaction should be monitored by TLC analysis. Suggested TLC solvent is 20% ethyl acetate in petroleum ether.) The pH of the reaction should remain 9—10. If necessary, additional sodium hydroxide solution should be added. Then the reaction is cooled to room temperature and the flask is further cooled in an ice-water bath to enforce crystallization. The product is filtered over suction, washed with 2 × 20 mL[①] of cold water, air-dried, and weighed.

The crude product is recrystallized by 95% ethanol to afford the pure product. If the solution is dark in color, activated carbon may be added to decolorize the recrystallization solution at high temperatures. TLC analysis and melting point analysis is conducted to characterize the product.

Submit the product to the instructor. Clean-up and check-out before you leave the lab.

Technique Checklist

1. A reaction over cooling and/or heating conditions.
2. Recrystallization.
3. Vacuum filtration.
4. TLC analysis.
5. Melting point measurement.

Pre-lab Discussion

1. Mechanism of the benzoin condensation.
2. TLC analysis and recrystallization.
3. Dropwise addition of one reagent to the other.

Experiment Outline

1. Mix vitamin B_1 to a flask containing water and 95% ethanol, and cool the mixture to ~4 ℃.

---

① 2 × 20 mL means to use 20 mL of the solvent/ solution for the same operation twice.

2. Dropwise add 10% NaOH(aq) to the mixture at ~4 ℃.

3. Allow the reaction to incubate at 60—75 ℃ for 1 h and then at ~90 ℃ for 0.5 h.

4. Cool the reaction to room temperature and further cool it to ~4 ℃ to stimulate precipitation of the product.

5. Perform recrystallization using 95% ethanol.

6. Perform TLC analysis during reaction and on the final product.

Helpful Hints

1. Vitamin B_1 in general is not very stable in the presence of base. Therefore, both vitamin B_1 and sodium hydroxide solution must be cooled to ~4 ℃ prior to mixing together to minimize the decomposition of vitamin B_1.

2. The reaction can take place at lower temperatures, such as room temperature. But it takes several days for the reaction to complete. Under current lab set-up, heating is required to accelerate the reaction.

3. It is important for the reaction mixture to slowly cool down to room temperature prior to the cooling to ~4 ℃. Otherwise, the product may crash out too fast to capsulate a large amount of impurities in the crystals.

4. The solubility of benzoin in 95% ethanol is ~12—14 g/100 mL while boiling.

Results

1. Record the phenomena during the experiment.

2. Record the TLC plates for reaction monitoring, product purity, and product identification.

3. Record the weights of the crude product and the recrystallized product.

4. Record the melting point of the product.

5. Record the appearance of the product.

6. Calculate the isolated yield of the reaction and R_f for the product.

5.8 Synthesis of Cinnamic Acid by the Perkin Reaction

Objectives

1. To understand the mechanism of the Perkin reaction.

2. To prepare cinnamic acid via a modified Perkin reaction using potassium carbonate as the base.

3. To practice azeotropic distillation, precipitation by pH adjustment, and recrystallization.

4. To perform TLC and melting point analysis.

Hazards

1. Chemical hazards

Benzaldehyde, acetic anhydride, potassium carbonate, conc. hydrochloric acid, water, 10% sodium hydroxide solution, activated carbon

2. Physical hazards

Heating might cause skin-burning and broken glassware may lead to skin damage.

Equipment

1. 250 mL Beaker.
2. Büchner funnel and filtration flask.
3. Condensers.
4. Congo red paper.
5. Distillation head and receiver adapter.
6. 100 mL Erlenmeyer flask.
7. Filter paper and rubber adapter.
8. 10 mL, 25 mL or 50 mL Graduated cylinder.
9. Heating set-up.
10. Iron ring, iron stand and clamps.
11. Liquid funnel and filter paper.
12. Melting point apparatus and capillary tubes for melting point test.
13. Rubber tubing.
14. 125 mL Separatory funnel with a grounded joint.
15. Stirring rod.
16. 5 mL Test tube.
17. Thermometer and adapter.
18. TLC plates and capillary tubes.
19. TLC chamber and aluminium foil.
20. 50 mL Two-neck round bottom flask.
21. UV light at 254 nm.
22. Water pump.

Pre-lab work

Before entering the wet lab, read thoroughly about the experiment and understand the physicochemical principle behind. Plan your set-up of equipment by visiting useful sources and complete your pre-lab report.

Introduction

The aldol reaction and aldol condensation are classic reactions between aldehydes to yield β-hydroxyaldehydes or α, β-unsaturated aldehydes. Under certain conditions, ketones can undergo such reactions if the product is quickly removed from the system. It is widely known that either an acid or a base can catalyze the reaction. In either case, enol or enolate is formed and functions as the nucleophile to attack the carbonyl in another molecule of aldehyde/ketone to give the β-hydroxyl aldehyde/ketone, which can quickly undergo dehydration under acidic conditions or upon heating to provide α, β-unsaturated aldehyde/ketone. There are many other similar reactions, using esters or amides as the enol/enolate precursors, such as the Claisen condensation, the Dieckmann condensation, and the Stobbe condensation for instance.

One of the similar condensation reactions uses acid anhydride containing α-hydrogen as the source of enol/enolate and aromatic aldehyde as the electrophile. This reaction, always referred as the Perkin reaction, requires the presence of alkali salts as the base catalyst, of which carboxylate salts and carbonate are frequently used. Such reaction generates β-aryl-α, β-unsaturated carboxylic acids, and was first developed by William Henry Perkin to make cinnamic acids. It has been reported by Kalnin that potassium carbonate accelerates the reaction with an improved yield. The reaction is indicated below:

$$C_6H_5CHO + (CH_3CO)_2O \xrightarrow[\Delta]{K_2CO_3} C_6H_5CH{=}CHCOOH + CH_3COOH$$

Cinnamic acid

Procedure

To a 50 mL two-neck round bottom flask fitted with a condenser and a thermometer are added 5 mL of fresh benzaldehyde, 7. 5 mL of fresh acetic anhydride and 3. 9 g of grounded anhydrous potassium carbonate. The mixture is then gently heated to reflux for 45 min. After being cooled back to room temperature, ~30 mL of 10% sodium hydroxide solution is added to the flask carefully to adjust the pH of the mixture to basic (pH ~8). A separatory funnel with ~45 mL of water is attached to the two-neck flask to replace the thermometer and the refluxing condenser is replaced by a distillation set-up. The reaction flask is heated for a steam distillation and when there are droplets collected in the receiving flask, the water in the separatory funnel is added in an approximate rate to keep the total liquid volume in the flask nearly unchanged. The steam distillation is terminated when no oil is further collected in the collection flask. The distillate is discarded into the organic non-halogen solvent waste, while the distillation flask is cooled to ~80 ℃ and ~3 g of activated carbon is added to the residue carefully. The mixture is further heated to reflux for 5 min, and allowed to cool down to room temperature. Gravity filtration is applied to separate the solid and liquid. After separation, two 10 mL water is added to wash the solid. All filtrate is combined in a 250 mL beaker and slowly added with conc. HCl to precipitate the product until the Congo red test paper turns blue. The product is collected by vacuum filtration, washed with 5 mL of dilute HCl and 5 mL of water, and dried in the air. The product should be analyzed by TLC using 90 : 10 (v/v) dichloromethane : methanol as the TLC developing solvent and melting point test. If necessary, the product is further purified by recrystallization with 5 : 1 water: ethanol co-solvent.

Submit the product to the instructor. Clean-up and check-out before leaving the lab.

Technique Checklist

1. Steam distillation.
2. Precipitation by pH adjustment.
3. Gravity filtration, vacuum filtration and recrystallization.
4. TLC analysis.
5. Melting point measurement.

Pre-lab Discussion

1. Mechanism of the Perkin reaction to prepare cinnamic acid.
2. Steam distillation and the set-up.
3. pH adjustment to change the solubility of a compound in the solution.

Experiment Outline

1. Heat benzaldehyde, Acetic anhydride, and potassium carbonate in a 50 mL two-neck round bottom flask to reflux for 45 min.
2. Add 10% sodium hydroxide to the mixture.
3. Perform steam distillation to remove benzaldehyde.
4. Add activation carbon to the residue at 80 ℃ and heat the mixture to reflux for 5 min.
5. Undertake gravity filtration is to remove the solid.
6. Add conc. HCl to precipitate the product at pH ~3.
7. Undertake vacuum filtration and recrystallization to purify the product.
8. Perform TLC and melting point analysis.

Helpful Hints

1. Heating is crucial for the reaction to proceed.

2. If potassium carbonate provided in the laboratory is not fine enough, the solid should be grounded with a mortar.

3. Steam distillation is useful to remove benzaldehyde at a temperature lower than its boiling point.

4. Sodium hydroxide facilitates the hydrolysis of acetic anhydride and ester.

5. Special attention should be given to the addition of *conc.* HCl to the reaction and this step is highly recommended to take place in the fumehood.

6. Congo red test paper is blue when the pH is lower than 3, but red when the pH is higher than 5.

7. The final product might undergo polymerization.

8. The melting point of cinnamic acid is 133 ℃.

Results

1. Record the phenomena during the experiment.
2. Record the mass of the crude product and recrystallized product.
3. Record the TLC result and melting point of the product.
4. Record the appearance of the product.
5. Calculate the isolated yield of the reaction and the R_f value of the product.

5.9 Synthesis of Cinnamic Acid by the Knoevenagel Condensation

Objectives

1. To understand the mechanism of the Knoevenagel reaction.
2. To practice extraction and recrystallization.
3. To perform TLC and melting point analysis.

Hazards

1. Chemical hazards

Benzaldehyde①, malonic acid, potassium carbonate, 2 mol · L^{-1} hydrochloric

① The other derivatives of benzaldehyde, such as *p*-methoxybenzaldehyde, *p*-nitrobenzaldehyde, or *p*-chlorobenzaldehyde can also be used for this experiment. It will be interesting to group students for different benzaldehyde precursors to compare the influence of substituents on the reaction rates and conversion yield.

acid, pyridine, piperidine.

2. Physical hazards

Heating might cause skin-burning and broken glassware may lead to skin damage.

Equipment

1. Büchner funnel and suction flask.
2. Condensers.
3. Filter paper and rubber adapter.
4. 10 mL, 25 mL or 50 mL Graduated cylinder.
5. Heating set-up.
6. Iron ring, iron stand and clamps.
7. Melting point apparatus and capillary tubes for melting point test.
8. 50 mL Round bottom flask.
9. Rubber tubing.
10. 50 mL Separatory funnel.
11. TLC plates and capillary tubes.
12. TLC chamber and aluminium foil.
13. UV light at 254 nm.
14. Water pump.

Pre-lab work

Before entering the wet lab, read thoroughly about the experiment and understand the physicochemical principle behind. Plan your set-up of equipment by visiting useful sources and complete your pre-lab report.

Introduction

Aldehydes or ketones can react with active methylene compounds, such as β-keto acids and malonic acid or derivatives to provide α, β-unsaturated dicarbonyl products under the catalysis of a weak base. This reaction is known as the Knoevenagel condensation and is an *aldol*-like reaction catalyzed by bases. The reaction mechanism highly depends on the type of base used in the reaction. The products may undergo hydrolysis or decarboxylation to give α, β-unsaturated carbonyl compounds. For instance, when treating benzaldehyde with malonic acid in the presence of pyridine and piperidine, the α, β-unsaturated dicarboxylic acid, 2-benzylidenemalonic acid first forms, and quickly decarboxylates to afford cinnamic acid as the final product upon heating.

CHO + HOOC COOH $\xrightarrow[\triangle]{\text{pyridine, piperidine}}$ COOH

To avoid competitive side reactions, aldehydes or ketones containing no α-hydrogen atoms are frequently used as the substrate. The active methylene compounds, possessing α-hydrogen atoms, are of a wide range, usually have two electron withdrawing groups, and include malonic esters, malonodinitrile, and acetoacetate esters, *etc*. The weak base as the catalyst can be primary, secondary, and tertiary amines as well as many inorganic compounds. The reaction can be accelerated by removing the side product, water, using measures of azeotropic distillation or addition of a dehydrating reagent such as molecular sieves.

Procedure

To a 50 mL separatory funnel are added 20 mL of water, 8 g of potassium carbonate, and 3. 6 mL of benzaldehyde. The mixture is mixed thoroughly and allowed to stand still for 0. 5 h until the layers separate. The aqueous layer is discarded and the organic layer is added to a pre-warmed mixture of 3. 1 g of malonic acid and 5 mL of pyridine at 60 ℃ in a 50 mL round bottom flask. After 10 drops of piperidine are added, the reaction mixture is heated to 80 ℃. When no bubbles form any more, the reaction mixture is added with around 40 mL of 2 mol • L^{-1} hydrochloric acid and then cooled to room temperature. The mixture is subject to suction filtration to collect the solid, which is further washed with 3 × 20 mL of 2 mol • L^{-1} hydrochloric acid, 3 × 20 mL of water, 3 × 10 mL of 20% aqueous ethanol, and 3×20 mL of light petroleum ether under suction. The product is further dried in an oven at 80 ℃.

The product should be analyzed by TLC using 90 : 10 (v/v) dichloromethane : methanol as the TLC developing solvent, and melting point test. If necessary, the product is further purified by recrystallization with 5 : 1 water : ethanol co-solvent.

Submit the product to the instructor. Clean-up and check-out before leaving the lab.

Technique Checklist

1. Extraction to remove acidic components.
2. Precipitation by pH adjustment.
3. Vacuum filtration and recrystallization.
4. TLC analysis and melting point measurement.

Pre-lab Discussion

1. Mechanism of the Knoevenagel condensation to prepare cinnamic acid.
2. Purpose of using potassium carbonate.

Experiment Outline

1. Heat benzaldehyde, malonic acid, pyridine, and piperidine in a 50 mL round bottom flask to 80 ℃ until no bubbles form.
2. Add 2 mol • L^{-1} HCl is added to quench the reaction and precipitate the product.
3. Perform filtration and recrystallization.
4. Perform TLC and melting point analysis.

Helpful Hints

1. Benzaldehyde stored for a long time is usually partially oxidized.
2. The final product might undergo polymerization.
3. The melting point of cinnamic acid is 133 ℃.

Results

1. Record the phenomena during your operation.
2. Record the amount of the crude product and recrystallized product.
3. Record the TLC result and melting point of the product.
4. Record the appearance of the product.
5. Calculate the isolated yield of the reaction and the R_f value of the product.

5.10 Synthesis of Methyl Orange

Objectives

1. To understand the mechanism of the diazo coupling reaction.
2. To practice recrystallization and vacuum filtration.
3. To understand how a pH indicator works.

Hazards

1. Chemical hazards

Sulfanilic acid, *N*,*N*-dimethylaniline, conc. HCl, sodium nitrite, acetic acid, 5% and 10% NaOH aqueous solutions, sodium chloride, brine, ethanol, diethyl ether.

2. Physical hazards

Heating might cause skin-burning and broken glassware may lead to skin damage.

Equipment

1. 50 mL & 100 mL Beakers.
2. Büchner funnel and filter paper.
3. Filtration flask and rubber adapter.
4. 25 mL & 10 mL Graduated cylinders.
5. Heating set-up.
6. Ice-water bath.
7. Iron stand and clamps.
8. Litmus test paper or pH test paper.
9. Pasteur pipettes.
10. Stir rod.
11. 10 mL and 5 mL Test tubes.
12. Water pump.

Pre-lab work

Before entering the wet lab, read thoroughly about the experiment and understand the physicochemical principle behind. Plan your set-up of equipment based on the procedure described herein and complete your pre-lab report.

Introduction

Methyl orange, also named as sodium 4-[(4-dimethylamino) phenylazo] benzenesulfonate, is a widely used pH indicator and demonstrates beautiful and clear color changes in the pH range of 3.1 to 4.4. As it changes color at this pH range (slightly acidic) for a sharper end point than the universal indicators, it is usually used in titrations of acids.

The preparation of methyl orange can be achieved by the diazo coupling reaction between the diazotized sulfanilic acid and *N*,*N*-dimethylaniline. However, diazonium salts are typically unstable and tend to detonation to release a large quantity of energy and nitrogen gas. It raises serious dangers to isolate the diazonium salts and instead they are usually prepared prior to their usage without purifications. The preparation of diazonium salts is always carried out by adding nitrous acid to the aniline compounds, in which nitrous acid is *in situ* prepared by treating sodium nitrite with a strong acid such as hydrochloric acid. The diazotization of aniline compounds is demonstrated below.

$$R\text{-}C_6H_4\text{-}NH_2 \xrightarrow{NaNO_2,\ HCl} R\text{-}C_6H_4\text{-}\overset{+}{N_2} + Cl^-$$

diazonium salt

The process relies on the formation of nitrosonium ion from sodium nitrite and the acid. The nitrosonium ion is a very good electrophile and the amino group of the aniline compounds spontaneously nucleophilically attacks the nitrosonium ion to construct a novel covalent bond between the two nitrogen atoms. Subsequently dehydration eventually leads to thc diazonium salts. The diazonium salt carrying a positive charge is very electron deficient. As a result, when a good nucleophile is present, such as *N*,*N*-dimethylaniline, the nucleophile and electrophile would react with each other to form a new covalent bond. Typically for an arene as the nucleophile, an electrophilic aromatic substitution takes place and a diazo compound is hence formed.

$$R\text{-}C_6H_4\text{-}\overset{+}{N_2} + C_6H_5\text{-}R' \longrightarrow R\text{-}C_6H_4\text{-}N{=}N\text{-}C_6H_4\text{-}R + H^+$$

Overall, the sulfanilic acid is first diatonized to afford the diazotized sulfanilic acid, which subsequently reacts with *N*,*N*-dimethylaniline to provide methyl orange. The synthetic route of methyl orange is summarized below.

$$NH_2\text{-}C_6H_4\text{-}SO_3H \xrightarrow{NaOH} NH_2\text{-}C_6H_4\text{-}SO_3Na \xrightarrow[HCl]{NaNO_2} [N{\equiv}N^+\text{-}C_6H_4\text{-}SO_3H]Cl^-$$

Sulfanilic acid

$$\xrightarrow[HOAc]{C_6H_5N(CH_3)_2} (CH_3)_2HN^+\text{-}C_6H_4\text{-}N{=}N\text{-}C_6H_4\text{-}SO_3H\ \ {}^-OAc \xrightarrow{NaOH}$$

$$(CH_3)_2N\text{-}C_6H_4\text{-}N{=}N\text{-}C_6H_4\text{-}SO_3Na$$

Methyl Orange

In a solution that is more basic than pH 4.4, methyl orange appears as the deprotonated anion in colors from yellow to orange. But under acidic conditions lower than pH 3.2, it exists as the red protonated cation. The presence of methyl orange in acidic conditions or basic conditions is demonstrated in the acid-base reactions below. As an acid-base indicator, it is usually prepared as an aqueous solution with a concentration of 0.01%.

SO_3H ... N=N ... $^{+}NH(CH_3)_2$ ⇌ [$(CH_3)_2N$—C_6H_4—N=N^{+}(H)—C_6H_4—SO_3H ↔ $(CH_3)_2N^{+}$=C_6H_4=N—NH—C_6H_4—SO_3H ; λ_{max}=520 nm red] $\underset{HCl}{\overset{NaOH}{\rightleftharpoons}}$ SO_3Na—C_6H_4—N=N—C_6H_4—$N(CH_3)_2$; λ_{max}=460 nm orange

Procedure

1. *Preparation of the diazotized sulfanilic acid*

To one 50 mL beaker are mixed 2.0 g of sulfanilic acid and 10 mL of 5% sodium hydroxide solution. The beaker is heated in a warm water bath to facilitate the complete dissolution of the solid. The homogeneous solution is allowed to cool down to room temperature. Then, 0.80 g of sodium nitrite is added to the solution and the mixture is stirred until all solid disappears. This mixture is dropwise added using a Pasteur pipette to a 100 mL beaker containing 13 mL of cold water (0~5 ℃) and 2.5 mL of conc. HCl. To avoid decomposition of the diazonium salt, the reaction temperature should remain below 5 ℃. The diazonium of sulfanilic acid should quickly crash out as a finely divided white precipitate. The suspension should maintain in the ice-water bath for at least 15 min before its use.

2. *Preparation of methyl orange*

In a 10 mL test tube, 1.3 mL of *N*, *N*-dimethylaniline and 1.0 mL of glacial acetic acid are mixed together thoroughly. The solution is added dropwise to the cooled suspension of diazotized sulfanilic acid in the 100 mL beaker while the suspension is stirred using a stir rod. In a few minutes, a red precipitate should form. The mixture is kept cooled in an ice-water bath and stirred vigorously for about 10~15 min to ensure the completion of the coupling reaction. To the beaker is added 15 mL of 10% sodium hydroxide solution slowly along stirring the suspension while the beaker remains in an ice-water bath. The mixture should be basic all the time.[①] After sodium hydroxide is added, the mixture is heated to boil for 10 to 15 min to

① Check the pH of the reaction solution constantly with Litmus or pH paper to make sure the solution is basic. If not, additional sodium hydroxide solution should be added.

dissolve most of the crude methyl orange. When all solid disappears, 5.0 g of NaCl is added to the solution. After all salt dissolves, the mixture is cooled down in an ice-water bath. The crystals are collected by vacuum filtration and washed with 20 mL of brine. To further purify the product, the filter cake along with the filter paper is transferred to a large beaker containing 75 mL of boiling water. The mixture is stirred constantly and heated at gentle boil for a few minutes. The filter paper is removed and the solution is allowed to cool to room temperature. Then the mixture is further cooled in an ice-water bath and the product is collected by vacuum filtration. The product is further washed with a small quantity of ethanol (~2 mL) and diethyl ether (~2 mL), allowed to dry a bit in the air, weighed, and submitted.

3. *Determining the color of the indicator at different pHs*

Dissolve 5 mg of the product in 1 mL of water to make a solution of 0.5%. 2 Drops of this solution is added to 0.5 mL of dilute HCl or dilute NaOH solution. Record the color of the resulting mixture(s).

Submit the product to the instructor. Clean-up and check-out before leaving the lab.

Technique Checklist

1. Manual stirring.
2. Reagent addition by Pasteur pipettes.
3. Vacuum filtration and recrystallization.
4. Cooling.
5. The pH measurement of a reaction.

Pre-lab Discussion

1. Reaction mechanism for the preparation of methyl orange.
2. Recrystallization and suction filtration.
3. Safety when working with diazotization reactions.

Experiment Outline

1. Operate the experiment by stirring the mixture and dropwise adding reagent by a Pasteur pipette.

2. Perform recrystallization and suction filtration.

3. Obtain the mass of methyl orange and identify its color at different pH conditions.

Helpful Hints

1. It is absolutely forbidden to dry the diazonium salts because of their high

explosive nature.

2. Sulfanilic acid is a zwitterionic molecule. It has poor solubility under acidic conditions but its solubility in water increases significantly under basic conditions. In the diazotization process, sulfanilic acid is first dissolved in a basic solution. When it is mixed with the acidic nitrosonium ion, the sulfanilic acid is precipitated out of solution as a finely solid, which is immediately diazotized.

3. The temperature to prepare the diazonium salt must be maintained below 5 ℃ to avoid the decomposition of diazonium salt.

4. Due to the presence of precipitate, the mixture must be stirred vigorously and thoroughly to facilitate the reaction.

5. It might be helpful to add a little bit of sodium hydroxide during the recrystallization of the final product.

6. The liquid waste containing the diazonium salt is added with 1 g of potassium iodide. The resulting mixture is stirred for 20 min before being disposed into the aqueous waste container.

Results

1. Record the reaction phenomena during the experiment.

2. Record the weight and appearance of the product.

3. Record the color change when methyl orange solution is added to basic solution or acidic solution.

4. Calculate the isolated yield of the reaction.

5.11 Synthesis of Ethyl Benzoate

Objectives①

1. To review the Fischer esterification.

2. To master the continuous removal of water produced in a reaction by the Dean-Stark apparatus.

3. To practice extraction and simple distillation.

Hazards

1. Chemical hazards

Ethanol②, benzoic acid, conc. H_2SO_4, *cyclo*hexane, 10% sodium carbonate,

① This experiment can also be used to prepare butyl benzoate.

② Butan-1-ol is used to replace ethanol for the synthesis of butyl benzoate.

diethyl ether, anhydrous calcium chloride, 25% EtOAc/petroleum ether

2. Physical hazards

Heating might cause skin-burning and broken glassware may lead to skin damage.

Equipment

1. 100 mL Beaker.
2. Boiling chips.
3. 25 mL Graduated cylinder.
4. Dean-Stark apparatus.
5. Distillation kit (distillation head, thermometer with adapter, Liebig condenser, receiver adapter).
6. 50 mL Erlenmeyer flasks.
7. Condenser.
8. Heating set-up.
9. Iron ring, iron stand and clamps.
10. pH test papers.
11. 100 mL & 50 mL Round bottom flasks.
12. Rubber tubing.
13. 50 mL or 125 mL Separatory funnel.
14. Stir rod.
15. TLC chamber and aluminium foil.
16. TLC plate and capillary tubes.
17. UV light at 254 nm.

Pre-lab work

Before entering the wet lab, read thoroughly about the experiment and understand the physicochemical principle behind. Plan your set-up of equipment by visiting useful sources and complete your pre-lab report.

Introduction

The Fischer esterification is very efficient to prepare esters from corresponding carboxylic acids and alcohols. However, the reaction would produce water and quickly reach an equilibrium. Along with the progression of the reaction, more product is produced and the reaction rate decreases accordingly. To drive the conversion forward, it is reasonable to remove the product(s) out of the reaction mixture. As water can be removed by azeotropic distillation with various water-immiscible and inert organic solvents, such as toluene and *cyclo*hexane, *etc.*, the reaction progress can be monitored by the collected amount of water. In order to collect or remove water, the Dean-Stark apparatus was developed. It is usually required to use the less dense water-immiscible solvent that can form azeotrope with water and the vapor is condensed by a condenser and collected in the side-arm of the Dean-Stark apparatus(Figure 5. 1). The upper, less-dense layer can flow back to the reactor while the water layer remains in the trap. When the water level reaches the branch position in the side-arm, the trap must be drained into a receiving flask. A typical Dean-Stark apparatus set-up is shown Figure 5. 1.

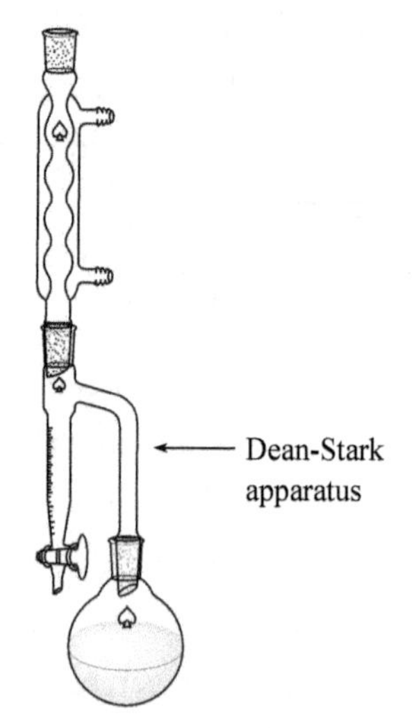

Figure 5. 1 Set-up for azetropic removal of newly produced water using a Dean-stark apparatus

Procedure

To a 100 mL round bottom flask are added 6. 1 g of benzoic acid, 13 mL of 95% ethanol, 10 mL of *cyclo*hexane, 2. 0 mL of sulfuric acid, and several boiling chips.① *Cyclo*hexane is added to fill the side-arm of the Dean-Stark apparatus which is attached to the round bottom flask. A condenser is connected to the Dean-stark apparatus and fixed with a clamp. Turn on the running water and the reaction flask is slowly heated to reflux in a water bath at ~70 ℃. Accompanied with the reaction progression, two layers appear in the side-arm of the Dean-Stark apparatus and the reaction completes until approximately 15 mL of the lower layer is collected. Then the solvent in the side-arm of the Dean-Stark apparatus is drained and all the volatiles (ethanol and *cyclo*hexane) are removed by adjusting the temperature of the water bath to 90 ℃. When all volatiles are removed, the residue is cooled a little bit and slowly poured into a 100 mL beaker containing 20 mL of cold water. The mixture is transferred to a separatory funnel and mixed thoroughly. The layers are separated and the aqueous layer is extracted with 10 mL of diethyl ether. The organic layers are combined, washed with 10 mL of 10% sodium carbonate and 10 mL of water respectively, and transferred to a dry Erlenmeyer flask for drying by anhydrous calcium chloride. The solid is filtered off and the filtrate is placed in a round bottom flask for simple distillation: Diethyl ether is removed under gently heating and then the flask is heated to collect the fraction of 211~213 ℃② in a dry and pre-weighed 50 mL Erlenmeyer flask. ~1 mg of the product and 1 mg of benzoic acid is dissolved

① For the synthesis of butyl benzoate, 11. 5 mL of butan-1-ol, 7. 3 g of benzoic acid, 10 mL of *cyclo*hexane, and 2 mL of sulfuric acid are used.

② Fraction of 249~252 ℃ is collected for butyl benzoate during distillation. Therefore, Vacuum distillation is applied for butyl benzoate.

in 0. 2 mL of ethanol respectively for TLC analysis using 25% EtOAc/petroleum ether as the TLC developing solvent.

Submit the product to the instructor. Clean-up and check-out before leaving the lab.

Technique Checklist

1. Azeotropic evaporation during a reaction.
2. Extraction and simple distillation.
3. TLC analysis.

Pre-lab Discussion

1. The reaction mechanism for the Fischer esterification.
2. Purpose of azeotropic evaporation and use of the Dean-Stark apparatus.
3. Extraction, distillation, and TLC analysis.

Experiment Outline

1. Heat the reaction mixture in a set-up with a Dean-Stark apparatus at 70 ℃.
2. Remove the volatile by heating the mixture at 90 ℃.
3. Perform extraction and simple distillation.
4. Perform TLC analysis.

Helpful Hints

1. The azeotropic boiling point for water-ethanol-cyclohexane is 62. 1 ℃.
2. The boiling point of ethyl benzoate is 211～213 ℃ at 1 atmosphere①.

Results

1. Record reaction phenomena along your operation.
2. Record the product mass and TLC result.
3. Record the appearance of the product.
4. Calculate the isolated yield of the reaction and the R_f of the product.

5.12 The Diels-Alder Reaction

Objectives

1. To use the Diels-Alder reaction to prepare alicyclic compounds.

---

① Boiling point of butyl benzoate is 250. 3 ℃.

2. To practice recrystallization and vacuum filtration.

Hazards

1. Chemical hazards

Xylenes, anthracene, maleic anhydride, dichloromethane, 10% EtOAc in petroleum ether.

2. Physical hazards

Heating might cause skin-burning and broken glassware may lead to skin damage.

Equipment

1. Büchner funnel and filtration flask.
2. Condenser.
3. 50 mL Erlenmeyer flask.
4. Filter paper and rubber adapter.
5. 50 mL Graduated cylinder.
6. Heating set-up.
7. Ice-water bath.
8. Iron stand and clamps.
9. Melting point apparatus and capillary tubes for melting point test.
10. 50 mL Round bottom flask.
11. Rubber tubing.
12. TLC chamber and aluminium foil.
13. TLC plate and capillary tubes.
14. UV light at 254 nm.
15. Water pump.

Pre-lab work

Before entering the wet lab, read thoroughly about the experiment and understand the physicochemical principle behind. Plan your set-up of equipment by visiting useful sources and complete the pre-lab report.

Introduction

The Diels-Alder reaction is one of the most useful reactions used in organic chemistry to construct cyclic compounds. It requires a diene and a dienophile to react in a [4+2] cycloaddition fashion and provides a product of a six-membered ring. The *diene* is a conjugate diene (1, 3-diene) containing 4 π electrons. The *dienophile*, on the other hand, typically contains a C—C π bond and would react with the diene. A good diene-dienophile pair usually is a reaction pair of an electron-rich diene and an electron-deficient double bond or triple bond. The reaction proceeds via a concerted mechanism, and is regarded as one type of pericyclic reactions, as shown below. Since the reaction is concerted, the stereochemistry of reactants and the symmetry of the molecular orbitals determine the stereochemistry of the products.

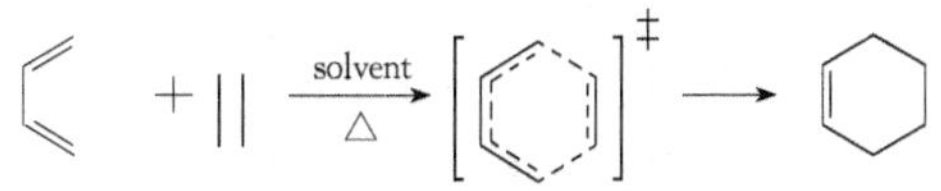

solvent
△
‡

In this experiment, maleic anhydride is used as the dienophile. The two strongly electron-withdrawing groups attached to the C—C double bond of the maleic anhydride make the double bond highly electron deficient and make maleic anhydride an excellent dienophile. The other reactant, anthracene, which is typically regarded as an aromatic compound, however has relatively lower aromaticity than isolated benzene rings. When the central ring of anthracene reacts as a diene, the product owns two isolated aromatic rings, which is shown below.

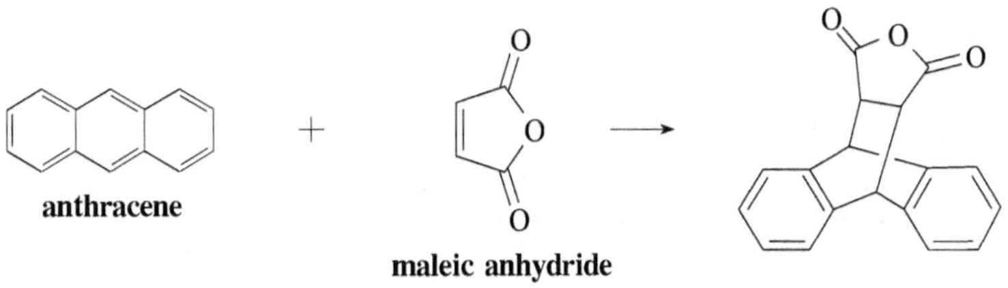

anthracene **maleic anhydride**

The Diels-Alder reaction usually requires high reaction temperatures. Accordingly, solvents with high boiling points are typically used, such as toluene and xylenes. In this experiment, xylenes, a mixture of the three isomers of dimethylbenzene, are used. This mixture has a boiling point of 140 ℃ and can provide a high reaction temperature upon heating. On the other hand, this solvent would not freeze when it is cooled in an ice-water bath. Moreover, the reactants of this reaction have a higher solubility in the solvent than the product while the product would easily crystallizes. The product is a relatively stable anhydride with a melting point of ~262—263 ℃. This compound may react with water in air slowly and is easily isolated and characterized before the hydrolysis occurs.

Procedure

To a 50 mL round bottom flask are added 2 g of anthracene, 1 g of maleic anhydride, and 25 mL of anhydrous xylenes. The mixture is heated at reflux for 30 min. Then the reaction mixture is cooled in an ice-water bath. The crude product is filtered in vacuum and washed with ~10 mL of pre-cooled xylenes. The crude product is weighed and then transferred to a 50 mL Erlenmeyer flask. The crude is further recrystallized by xylenes. [①] Allow the solution to cool to room temperature and then further cool the solution in an ice-water bath for 5 min. The solid is collected by vacuum filtration and rinsed with ~3 mL of cold xylenes. Let the product dry in the air for ~30 min prior to the measurement of melting point. Dissolve ~1 mg of the product, ~1 mg of anthracene, and ~1 mg of maleic anhydride in ~0.5 mL of dichloromethane respectively and conduct TLC analysis[②] with 10% EtOAc in petroleum ether. After the TLC development, the plate should be visualized under UV light at 254 nm. The melting point of the product should be measured.

Submit the product to the instructor. Clean-up and check-out before leaving the lab.

Technique Checklist

1. Refluxing.
2. Vacuum filtration.
3. Recrystallization.
4. TLC analysis.
5. Measurement of melting point.

Pre-lab Discussion

The mechanism of the Diels-Alder reaction.

Experiment Outline

1. Carry out a reaction under refluxing.
2. Perform vacuum filtration and recrystallization.
3. Measure the melting point.
4. Perform the TLC analysis.

Helpful Hints

1. The product should be dry before the melting point measurement.

2. Both the reaction and the recrystallization step should have a condenser attached.

3. During the reaction progression, if there is crystal on the inner wall of the reaction flask, try to rinse it down using the reaction solution by gently swirling the reaction flask.

---

① To carry out the recrystallization: First add ~5 mL of xylenes and heat the mixture gently to reflux. Slowly add more xylenes until all product dissolves.

② To achieve good visualization, it is best to dry out the loaded TLC plate before the TLC development as xylenes are also UV active and can absorb the UV light at 254 nm.

4. The melting point of the product is 262～263 ℃.

Results

1. Record the phenomena during the experiment.
2. Record the mass and appearance of the product.
3. Record the product's melting point and the TLC result.
4. Calculate the isolated yield of the reaction and the R_f of the product.

5.13 Synthesis of Acetanilide

Objectives

1. To undertake acylation reaction by a dehydration process.
2. To use fractional distillation.
3. To practice recrystallization and vacuum filtration.

Hazards

1. Chemical hazards

Aniline, glacial acetic acid, zinc powder[①], water, 10% methanol/dichloromethane.

2. Physical hazards

Heating might cause skin-burning and broken glassware may lead to skin damage.

Equipment

1. 500 mL Beaker.
2. Boiling chips.
3. Büchner funnel and filtration flask.
4. Condenser.
5. Distillation head.
6. Filter paper and rubber adapter.
7. 25 mL Graduated cylinder.
8. Heating set-up.
9. Iron stand and clamps.
10. Melting point apparatus and capillary tubes for melting point test.
11. Receiver adapter.
12. 50 mL Round bottom flask.

① A small zinc chunk can be used instead due to the safety regulation of zinc dust.

13. Rubber tubing.
14. 15 mL Test tubes.
15. TLC chamber and aluminium foil.
16. TLC plate and capillary tubes.
17. Vigreux distillation column.
18. UV light at 254 nm.
19. Water pump.

Pre-lab work

Before entering the wet lab, read thoroughly about the experiment and understand the physicochemical principle behind. Plan your set-up of equipment by visiting useful sources and complete the pre-lab report.

Introduction

Acylation of amines is a common reaction to construct an amide bond, which is present in many key natural compounds. In certain cases, acylation of amines can be used to protect the amino group in a molecule, while the acyl groups would be removed via hydrolysis at the end of a synthetic route. The acylation is usually undertaken with activated acyl reagents such as acyl halides or acid anhydrides. But these reagents are usually expensive and very reactive while the related acylation may be companied with bis-acylation. On the other hand, under certain conditions, carboxylic acid can be used for the acylation of amines. Although the acylation with corresponding carboxylic acid is slow, the reagent is very cheap. For instance, the acetylation of aniline by acetic acid as shown below can provide a relatively good yield within several hours if the side product is removed once it is formed to allow the reaction to go forward. The side product, water, can be removed by fractional distillation during the reaction progression.

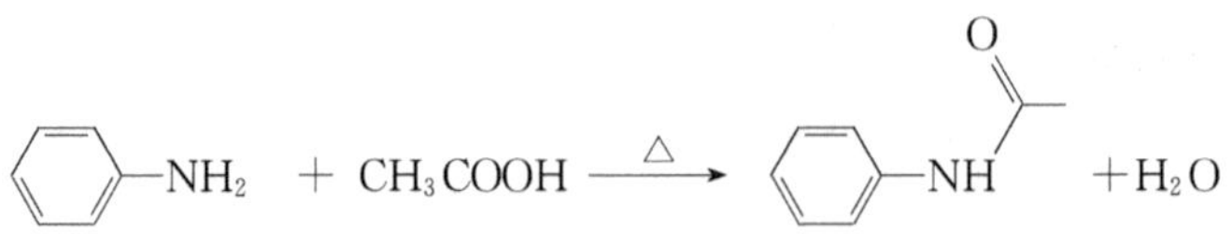

$$C_6H_5\text{-}NH_2 + CH_3COOH \xrightarrow{\triangle} C_6H_5\text{-}NHCOCH_3 + H_2O$$

Acetanilide

Procedure

To a 50 mL round bottom flask are added 10. 0 mL of freshly distilled aniline (0. 11 mol), 15. 0 mL of glacial acetic acid, and ~0. 1 g of zinc dust. The round bottom flask is then attached with a Vigreux distillation column, distillation head, thermometer, condenser, and receiver adapter for fractional distillation. The round bottom flask is heated to reflux and the temperature readout of the thermometer is kept at ~105 ℃ for around 1 hour. If a rapid temperature drop takes place, the reaction is complete and the reaction mixture should be cooled a bit and then poured to 250 mL of cold water. Until the mixture is cooled to room temperature, the mixture is filtered under vacuum and ~20 mL of cold water is used to thoroughly wash the crude. The crude product is then further purified by recrystallization with water. The solution is filtered to remove the zinc dust as soon as the crude dissolves and the filtrate is cooled to room temperature for crystallization. The product is collected by suction filtration, dried, and weighed. Dissolve ~1 mg of the product and ~1 mg of aniline in ~0. 2 mL of ethanol respectively for TLC analysis using 5% MeOH/DCM as the TLC developing solvent. After the TLC development, the plate should be visualized under light at 254 nm. The melting point of the product should be measured.

Submit the product to the instructor. Clean-up and check-out before leaving the lab.

Technique Checklist

1. Fractional distillation.
2. Recrystallization.
3. Vacuum filtration.
4. TLC analysis.
5. Melting point measurement.

Pre-lab Discussion

1. Principle of fractional distillation.
2. The mechanism for the synthesis of acetanilide.

Experiment Outline

1. Set up the fractional distillation for the reaction.
2. Pour the reaction to water and undertake suction filtration.
3. Perform recrystallization and suction filtration.
4. Obtain the mass of acetanilide.
5. Analyze the purity of the product by TLC and melting point analysis.

Helpful Hints

1. Make sure all joints are sealed well.
2. Carefully control the heating power to maintain the temperature as required.
3. Zinc dust is added to avoid the oxidation of aniline.

4. The reaction mixture should be poured to water while it is still hot. This is to avoid the potential precipitation of the product at lower temperature and difficulty of transferring the complete product out of the reaction vessel.

5. The solubility of acetanilide in 100 mL water is 5. 55 g at 100 ℃, 3. 45 g at 80 ℃, 0. 84 g at 50 ℃, and 0. 46 g at 20 ℃.

6. The melting point of acetanilide is 114 ℃. Therefore, try to avoid "oiling out" during the recrystallization.

Results

1. Record the reaction phenomena along the reaction.
2. Record the weight and appearance of the product.
3. Record the melting point of the product and the TLC result.
4. Calculate the isolated yield of the reaction and the R_f of the product.

5.14 Bromination of Acetanilide

Objectives

1. To *in situ* produce bromine by oxidation of bromide with bromate salt under acidic conditions.
2. To perform an electrophilic aromatic substitution.

Hazards

1. Chemical hazards

Potassium bromate, 48% hydrobromic acid in water, acetanilide, glacial acetic acid, 95% ethanol, 10% sodium bisulfite.

2. Physical hazards

Heating might cause skin-burning and needles or broken glassware may lead to skin damage.

Equipment

1. Büchner funnel and filtration flask.
2. Condenser.
3. 2 mL Disposable syringe with a detachable needle.
4. 50 mL Erlenmeyer flask.
5. Filter paper and rubber adapter.
6. 10 mL Graduated cylinder.
7. Heating set-up.

8. Iron stand and clamps.
9. Magnetic stirrer with a suitable magnetic stir bar.
10. Melting point apparatus and capillary tubes for melting point test.
11. Rubber tubing.
12. TLC chamber and aluminium foil.
13. TLC plate and capillary tubes.
14. UV light at 254 nm.
15. Water pump.

Pre-lab work

Before entering the wet lab, read thoroughly about the experiment and understand the physicochemical principle behind. Plan your set-up of equipment by visiting useful sources and complete the pre-lab report.

Introduction

Electrophilic aromatic substitution reactions are useful to introduce various functional groups onto the benzene ring. Bromination is one of the reactions that functionalize benzene and produces bromobenzene, which can be further used in reactions such as Suzuki coupling reactions. However, bromine (Br_2) frequently used as the brominating reagent is very corrosive and toxic. There have been a lot of efforts to produce bromine in the reaction in order to avoid and decrease the bromine directly handled in operation. For instance, bromate (BrO_3^-) can oxidize bromide (Br^-) under acidic conditions to provide Br_2 with high efficiency. To use the bromine (Br_2) produced by this procedure, the bromination reactions must not be water or acid sensitive.

$$BrO_3^- + 5Br^- + 6H^+ \longrightarrow 3Br_2 + 3H_2O$$

Many electrophilic aromatic substitution reactions require Lewis acids, such as BF_3, $FeCl_3$, or $AlCl_3$ to further activate the electrophiles. Unfortunately, most of such Lewis acids cannot stand protic conditions. On the other hand, aromatic compounds can be activated by electron-donating substituents and usually are active enough to nucleophilically attack electrophiles in protic solvent. For instance, the bromination of phenol in either CCl_4 or H_2O can afford electrophilic aromatic substituted products without the participation of Lewis acid catalysts. Acetanilide, also called *N*-acetyl aniline, contains the electron-donating group acetylamino-group, which is less electron-donating than the amino group, can be a good substrate for bromination in combination with bromate and bromide. The reaction is summarized below.

$$\text{acetanilide} \xrightarrow[\text{HOAc, } H_2O]{\text{HBr, } KBrO_3} \text{4-bromoacetanilide}$$

Procedure[①]

Weigh 1.0 g of acetanilide and 0.43 g of potassium bromate into an Erlenmeyer flask equipped with a stir bar. To the solid is added 10 mL of glacial acetic acid. The mixture is stirred rapidly and 1.5 mL of 48% hydrobromic acid is added slowly using a syringe.[②] Then the reaction is allowed to stir at room temperature for 30 min and poured to 150 mL of ice-cold water. The mixture is continuously stirred for another 15 min till all ice melts and vacuum filtration is applied to collect the solid. The solid is subsequently washed with 20 mL of 10% sodium bisulfite and 20 mL of cold water. The solid is further purified by recrystallization with 95% ethanol and the crystals are collected by vacuum filtration and air-dried.

The product should be characterized by melting point measurement and TLC analysis. For TLC analysis, dissolve ~2 mg of the product in 0.2 mL of ethanol and load on to a TLC plate. Then the plate is developed using 1 : 1 EtOAc : petroleum ether as the developing solvent. After TLC development, the plate should be visualized under UV light at 254 nm.

Submit the product to the instructor. Clean-up and check-out before leaving the lab.

Technique Checklist

1. Magnetic stirring for a reaction.
2. Reagent transfer using a syringe.
3. Vacuum filtration and recrystallization.
4. TLC analysis and measurement of melting point.
5. Quenching reactive reagents.

Pre-lab Discussion

1. Mechanism of the electrophilic aromatic substitution.
2. Purpose to use sodium bisulfite.

---

① The procedure is modified from Paul F. Schatz, Bromination of Acetanilide. *Journal of Chemical Education*, 1996, 73(3), 267.

② Do deal with hydrobromic acid in the hood with appropriate personal protection, such as gloves and googles. After addition, the syringe should be quenched with sodium bisulfide before disposal. The needle should be disposed separately into a sharp waste container after being quenched.

Experiment Outline

1. Add 48% hydrobromic acid to the mixture of other reagents using a syringe at room temperature.
2. Magnetically stir the reaction for 0.5 h.
3. Pour the reaction to an ice-water mixture.
4. Perform vacuum filtration and recrystallization.
5. Analyze the purity of the product by TLC and melting point test.

Helpful Hints

1. The first filtrate and washing solution should be quenched by sodium bisulfide.
2. The melting point of 4-bromoacetanilide is 170~171 ℃.

Results

1. Record the reaction phenomena.
2. Record the mass and appearance of the product.
3. Record the melting point and TLC result.
4. Calculate the isolated yield and the R_f of the product.

5.15 Deprotection of the Acetyl Protecting Group from *p*-Bromoacetanilide

Objectives

1. To remove acetyl group under basic conditions.
2. To use vacuum filtration and recrystallization.

Hazards

1. Chemical hazards

Ethanol, *p*-bromoacetanilide, potassium hydroxide, water, ethyl acetate, petroleum ether, methanol, dichloromethane.

2. Physical hazards

Heating might cause skin-burning and broken glassware may lead to skin damage.

Equipment

1. Büchner funnel and filtration flask.

2. Condenser.
3. 100 mL Erlenmeyer flask.
4. Filter paper and rubber adapter.
5. 10 mL Graduated cylinder.
6. Heating set-up.
7. Iron stand and clamps.
8. Melting point apparatus and capillary tubes for melting point test.
9. 50 mL Round bottom flask.
10. Rubber tubing.
11. Stir rod.
12. TLC chamber and aluminium foil.
13. TLC plate and capillary tubes.
14. UV light at 254 nm.
15. Water pump.

Pre-lab work

Before entering the wet lab, read thoroughly about the experiment and understand the physicochemical principle behind. Plan your set-up of equipment by visiting useful sources and complete the pre-lab report.

Introduction

Amide bonds are very important in organic chemistry. Some amides can be synthesized from the typical acyl nucleophilic substitution reactions between an amine and a carboxylic derivative, such as ester, acyl halide, anhydride, or carboxylic acid. On the other hand, in certain cases, it is used to block the functionality of amino groups in order to improve the regioselectivity and chemoselectivity, because amino groups are not only great nucleophiles and bases, but also demonstrate extraordinary electron donating effects. So the amino groups can be masked with simple acyl groups through amide bond formation. Right after the desired reaction takes place, the acyl group is mandatorily removed to reveal the amino group's reactivity for further manipulation. Acetyl group is one of the most used to protect amines by treating the amine with acetyl chloride or acetic anhydride with high efficiency. Due to the inertness of the amide bond under acidic or basic conditions, the hydrolytic cleavage of acetamide bond is typically under a harsh environment, such as in hot strongly acidic or basic solutions. Introduction of electro-withdrawing groups to the acetyl group can increase the lability of such hydrolysis.

Deprotection of the acetyl groups from acetamides can be achieved under various conditions, such as refluxing in acidic conditions or basic conditions.[①] The corresponding reactions are shown below.

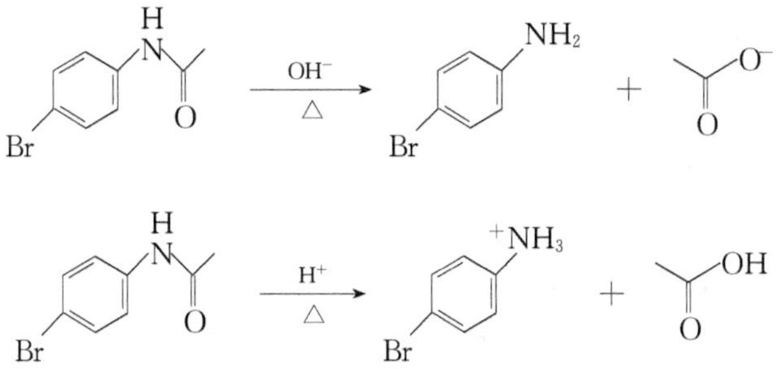

Procedure for the bromination

To a 50 mL round bottom flask are added 3.2 g of *p*-bromoacetanilide and 10 mL of ethanol. The mixture is carefully mixed to give a homogeneous solution and then is added with 3.0 g of potassium hydroxide dissolved in 5 mL of water. The resulting mixture is heated to reflux for 2 hours with a condenser attached to the round bottom flask. The solution is cooled to room temperature and poured into 30 mL of ice-cold water. The aqueous suspension is thoroughly mixed using a stir rod until the complete precipitation of the solid. The solid is collected via vacuum filtration and further washed with a small amount of ice cold water (~5 mL). The crude is weighed and recrystallized with 50% aqueous ethanol. The crystals are collected by filtration and air-dried.

The product should be characterized by melting point measurement and TLC analysis. For TLC analysis, dissolve ~2 mg of the product in 0.2 mL of ethanol and load the sample on to a TLC plate. Then the plate is developed using 1 : 1 EtOAc : petroleum ether and 1 : 9 MeOH : CH_2Cl_2 as the developing solvent. After TLC development, the plate should be visualized under UV light at 254 nm.

Submit the product to the instructor. Clean-up and check-out before leaving the lab.

Technique Checklist

1. Refluxing.
2. Vacuum filtration.
3. Recrystallization.
4. TLC analysis and measurement of melting point.

---

① Peter G. M. Wuts, Theodora W. Greene, *Protective Groups in Organic Synthesis*, 4ed., Wiley-Interscience, 2006.

Pre-lab Discussion

Mechanism of the deprotection of acetyl groups.

Experiment Outline

1. Heat the reaction at refluxing for 2 h.
2. Precipitate the product from water.
3. Perform vacuum filtration.
4. Recrystallize the crude using 50% aqueous ethanol.
5. Analyze the purity of the product by TLC and melting point test.

Helpful Hints

1. When using water to precipitate the product, more water should be added if necessary.

2. The melting point of 4-bromoacetanilide is 61～63 ℃.

Results

1. Record the reaction phenomena.
2. Record the mass and appearance of the product.
3. Record the melting point and TLC result.
4. Calculate the isolated yield and the R_f of the product.

5.16 Synthesis of Aspirin

Objectives

1. To synthesize aspirin by acetylation of salicylic acid.
2. To practice vacuum filtration and recrystallization.

Hazards

1. Chemical hazards

Salicylic acid, acetic anhydride, conc. sulfuric acid, water, saturated sodium carbonate solution, conc. hydrochloric acid, ethyl acetate, petroleum ether, methanol, dichloromethane, ethanol.

2. Physical hazards

Heating might cause skin-burning and broken glassware may lead to skin damage.

Equipment

1. 100 mL Beaker.
2. Büchner funnel and suction flask.
3. Condenser.
4. 10 mL and 50 mL Graduate cylinders.
5. Filter paper and rubber adapter.
6. Heating set-up.
7. Ice-water bath.
8. Iron stand and clamps.
9. Melting point apparatus and capillary tubes for melting point test.
10. pH test paper.
11. 50 mL Round bottom flask.
12. Rubber tubing.
13. Spatula.
14. Test tubes.
15. TLC chamber and aluminium foil.
16. TLC plate and capillary tube.
17. UV light at 254 nm.
18. Water pump.

Pre-lab work

Before entering the wet lab, read thoroughly about the experiment and understand the physicochemical principle behind. Plan your set-up of equipment by visiting useful sources and complete the pre-lab report.

Introduction

Aspirin, also referred as acetylsalicylic acid, is widely used for alleviating pains and treating a fever. It was first isolated in willow barks and found to be the key component responsible for the biological activity of willow barks used in ancient medicine. It has also been found recently that aspirin can be a treatment for immune diseases, prevent platelet aggregation, reduce cardiovascular diseases, intercept the β-amyloid formation to prevent or slow down the progression of the Alzheimer's disease, and treat diabetes and intestinal tumors. This organic compound was first synthesized in the Baeyer company by treating salicylic acid with excessive acetic anhydride, the reaction of which is demonstrated below.

COOH OH $\xrightarrow{(CH_3CO)_2O}$ COOH O O CH_3

The reaction works better when a small quantity of strong acid is used as the catalyst, such as phosphoric acid or sulfuric acid. This is because the acid can destroy the intramolecular hydrogen bonding for salicylic acid, which would allow the reaction to proceed at lower temperatures and decrease the extent of the side reaction for poly (salicyclic acid).

Procedure for the acetylation of salicylic acid

To a dry round bottom flask is added 4 g of salicylic acid, 10 mL of acetic anhydride, and 5 drops of conc. sulfuric acid①. After a condenser is attached to the flask, it is incubated in a boiling water bath for 20 min. Then 4 mL of water is added to the flask slowly to quench the reaction by the decomposition of acetic anhydride and the resulting mixture is further incubated in the boiling water bath for 2 min. After the boiling water bath is removed, 30 mL of water is added to the mixture and the flask is allowed to cool to room temperature, and further over an ice-water bath. The mixture is filtered in vacuum and the solid collected is further washed with ~10 mL chilled water. The solid is transferred to a 100 mL beaker and 10 mL of water is added. While the mixture is stirred, saturated sodium carbonate solution is added until no further gas is released and the pH of the obtained solution is ~8. The solution is filtered to remove any insoluble component and the insoluble particles are further washed with ~10 mL of water. The filtrate is combined and concentrated hydrochloric acid is added carefully to adjust the pH of the solution to 2 and the resulting mixture is further cooled in an ice-water bath for 5 min. The solid is then collected by suction filtration, washed with ~5 mL of cold water, and air dried. The solid is further recrystallized by ethyl acetate and the crystal obtained is dried in air and weighed.

The product should be characterized by melting point measurement and TLC analysis. For TLC analysis, dissolve ~2 mg of the product and ~2 mg of salicylic acid in 0.2 mL of ethanol respectively and load the samples on to a TLC plate. Then the plate is developed using 1 : 1 EtOAc : petroleum ether and 1 : 9 MeOH : CH_2Cl_2 as the developing solvent, respectively. After TLC development, the plate should be visualized under UV light at 254 nm.

---

① Pyridine, sodium acetate and/or trifluoroborane etherate can also be used as the catalyst for the reaction. It would be worthwhile to compare the reaction rates of the acetylation in presence of different catalyst by measuring the time it takes for the reaction's temperature to increase for 4 ℃.

Ferric test: 0.5 mL of the product alcoholic solution is added with 5 drops of 1% ferric chloride in ethanol for a characteristic test.

Spectroscopy characterization as UV-Visible absorption spectroscopy, infrared spectroscopy, and NMR spectroscopy is recommended for the final product.

Submit the product to the instructor. Clean-up and check-out before leaving the lab.

Technique Checklist

1. Reaction heated in a boiling water bath.
2. Suction filtration.
3. Neutralization and acidification.
4. Recrystallization.
5. TLC analysis and measurement of melting point.

Pre-lab Discussion

1. Mechanism of the acetylation of salicylic acid for aspirin.
2. Principle for the ferric test.

Experiment Outline

1. Slowly add sulfuric acid to the reaction.
2. Heat the reaction in a boiling water bath for 20 min.
3. Quench the reaction and precipitate the product with water.
4. Suction filtration.
5. Convert the product into salt by saturated sodium carbonate.
6. Suction filtration to remove insoluble components.
7. Acidification of the filtrate and suction filtration.
8. Recrystallization to further purify the product.
9. TLC analysis and melting point measurement of the product.
10. Ferric test for the product and precursor.

Helpful Hints

1. All glassware should be dry for the acetylation.

2. Acetic anhydride is volatile and its measurement should be undertaken in a hood.

3. All chemicals should be dissolved before the mixture is heated in the water bath.

4. The most possible by-product is poly(salicylic acid) and it has a poor solubility in water under either acidic conditions or basic conditions.

5. The addition of either sodium carbonate solution or conc. hydrochloric acid should be slow and careful to prevent loss of the desired product.

6. Phenol forms colored complex with ferric.

7. The melting point of aspirin is 138～140 ℃.

Results

1. Record the reaction phenomena.
2. Record the mass, melting point, and TLC result.
3. Calculate the isolated yield of the reaction.
4. Calculate the R_f for the product.
5. Record the phenomena of the ferric test.

5.17 Synthesis of Ferrocene via the Phase Transfer Catalysis

Objectives

1. To prepare ferrocene via *cyclo*penta-1,3-diene.
2. To apply phase transfer catalysis in the reaction.

Hazards

1. Chemical hazards

Tetrahydrofuran, *cyclo*penta-1, 3-diene①, 18-crown-6, potassium hydroxide, ferrous chloride tetrahydrate, conc. hydrochloric acid, ice, water, hexane, ethyl acetate, petroleum ether.

2. Physical hazards

Broken glassware may lead to skin damage.

Equipment

1. 100 mL Beaker.
2. Büchner funnel and suction flask.
3. 50 mL Erlenmeyer flask.
4. Filter paper and rubber adapter.
5. Glass rod.
6. 50 mL Graduated cylinder.
7. Iron stand and clamps.

---

① It should be freshly cracked before the experiment by the instructor or lab technician.

8. Melting point apparatus and capillary tubes for melting point test.
9. Magnetic stirrer and a suitable magnetic stir bar.
10. Test tubes.
11. TLC chamber and aluminium foil.
12. TLC plate and capillary tube.
13. UV light at 254 nm.
14. Water pump.

Pre-lab work

Before entering the wet lab, read thoroughly about the experiment and understand the physicochemical principle behind. Plan your set-up of equipment by visiting useful sources and complete the pre-lab report.

Introduction

Ferrocene is a diamagnetic orange or yellow coordinate complex of Fe(II). Ferrocene was found to hold a sandwich structure while the ligand *cyclo*pentadienyl ring, filled with 6π electrons, is aromatic. Ferrocene displays high stability at room temperature, and it is inert in catalytic hydrogenation or Diels-Alder reactions. It has also been found that ferrocene can be isolated via sublimation, which demonstrates its high stability even at high temperatures.

Typically, ferrocene is prepared from ferrous and the conjugate base of *cyclo*penta-1, 3-diene. The conjugate base can be easily obtained from the deprotonation of *cyclo*penta-1, 3-diene with a base, such as sodium hydroxide or potassium hydroxide. Then it quickly coordinates with ferrous cation in a ratio of 2:1. However, either sodium hydroxide or potassium hydroxide is hygroscopic and may not efficiently contact and react with *cyclo*penta-1,3-diene so that the reaction is usually undertaken in an inert and anhydrous atmosphere, which requires either nitrogen/argon supply or a glove box in the laboratory.

On the other hand, if water is present in the reaction vessel, phase transfer reagents can help the inorganic base to enter the organic solvent and hence accelerate the reaction significantly. For instance, 18-crown-6, one of the most used crown ethers, can trap potassium cations and hence increase the solubility of potassium hydroxide in organic solvents.

$$2\,C_5H_6 + FeCl_2 \xrightarrow[\text{18-crown-6, THF}]{\text{KOH}} Fe(C_5H_5)_2$$

Procedure for ferrocene preparation[①]

To a 50 mL Erlenmeyer flask with a magnetic stir bar is added 1 g of *cyclo*pentadiene, 2 g of 18-crown-6, and 30 mL of tetrahydrofuran (THF). Then 1.25 g of potassium hydroxide is added to the solution and the resulting mixture is then stirred at room temperature for 15 min and 1.5 g of ferrous chloride tetrahydrate is slowly added to the solution in several portions over 5 min. The reaction is further stirred vigorously for 35 min. The reaction mixture is carefully poured into 4 mL of conc. HCl with 20 g of ice in a 100 mL beaker and the Erlenmeyer flask is rinsed with 10 mL cold water. All solutions are combined and stirred until no visible ice is present. The product is collected by suction filtration and washed with several portions of 15 mL of water. The product is further purified via recrystallization using hexane.

The product should be characterized by melting point measurement and TLC analysis. For TLC analysis, dissolve ~5 mg of the product in 1 mL of ethyl acetate and load the sample solution onto a TLC plate. Then the plate is developed using 1 : 4, 1 : 1, and 4 : 1 ethyl acetate : petroleum ether as the developing solvents respectively. After the TLC development, the plate should be visualized under UV light at 254 nm.

Submit the product to the instructor. Clean-up and check-out before leaving the lab.

Technique Checklist

1. Suction filtration.
2. Recrystallization.
3. Magnetic stirring.
4. TLC analysis and measurement of melting point.

Pre-lab Discussion

1. Mechanism of the ferrocene preparation.
2. Origin of the stability of ferrocene.
3. Principle of phase transfer catalysis.

Experiment Outline

1. Deprotonation of cyclopentadiene in presence of a phase transfer reagent.

---

① This procedure is a modification from Sališová, M., Alper, H., phase Transfer Catalyzed Synthesis of Ferrocene Derivative, *Angewandte Chemie International Edition*. 1979, *18*(*10*), 792.

2. Coordination with ferrous over stirring conditions.

3. Suction filtration.

4. Recrystallization to further purify the product.

5. Analyze the purity of the product by TLC analysis and melting point test.

Helpful Hints

1. When using water to precipitate the product, more water should be added if necessary.

2. The melting point of ferrocene is 173～174 ℃.

Results

1. Record the reaction phenomena.

2. Record the mass, melting point, and TLC result.

3. Calculate the isolated yield for the reaction and the R_f for the product.

5.18 Synthesis of 2-Nitro-1,3-benzenediol

Objectives

1. To undertake steam distillation for thermosensitive compounds.

2. To understand the orientation rule for electrophilic aromatic substitution (EAS) and apply EAS in experiments.

3. To appreciate protection and deprotection in a synthetic route.

4. To practice recrystallization.

Hazards

1. Chemical hazards

1,3-Dihydroxybenzene, conc. sulfuric acid, 65% ～ 68% nitric acid, urea, ethanol, water, 20% ethyl acetate in petroleum ether, ice

2. Physical hazards

Heating might cause skin-burning and broken glassware may lead to skin damage.

Equipment

1. 100 mL Beaker.

2. Büchner funnel and suction flask.

3. Condenser.

4. Distillation head and receiver adapter.

5. 25 mL and 100 mL Erlenmeyer flasks.
6. Filter paper and rubber adapter.
7. 10 mL and 25 mL Graduated cylinders.
8. Heating set-up.
9. Ice-water bath.
10. Iron stand and clamps.
11. Magnetic stirrer and magnetic stir bar.
12. Melting temperature apparatus and capillary tube.
13. Mortar and pestle.
14. Pasteur pipettes.
15. Rubber septum and glass tubes.
16. Rubber tubing.
17. Separatory funnel with a ground joint.
18. Stirring rod.
19. Test tubes.
20. Thermometer with adapter.
21. TLC plate and capillary tube.
22. TLC chamber and aluminium foil.
23. 100 mL Two-neck round-bottom flask.
24. UV light at 254 nm.
25. Water bath.
26. Water pump.

Pre-lab work

Before entering the wet lab, read thoroughly about the experiment and understand the physicochemical principle behind. Plan your set-up of equipment by visiting useful sources and complete the pre-lab report.

Introduction

Electrophilic aromatic substitution (EAS) is an important reaction to functionalize aromatic compounds. The reaction highly relies on the substituents attached on the aromatic system. In fact, different substituents demonstrate distinct orientation effects. For instance, the electron-donating group (EDG) usually guides the reaction to provide products at its *ortho*- or *para*-position, while the electron-withdrawing group (EWG) provides products at its *meta*-position. However, when two or more substituents are present on the benzene ring, the EAS product will be determined by the stereoelectronic effects of the substituents. But in general, the strong EDG will be determinative to the orientation of the newly introduced substituent in the product. To obtain 2-nitro-1,3-benzenediol, a nitration reaction can take place. However, multiple products can be produced while 2-nitro-1, 3-benzenediol is the minor product due to the steric hindrance raised by the substituents at 1- and 3-positions. The corresponding nitration reaction is demonstrated below.

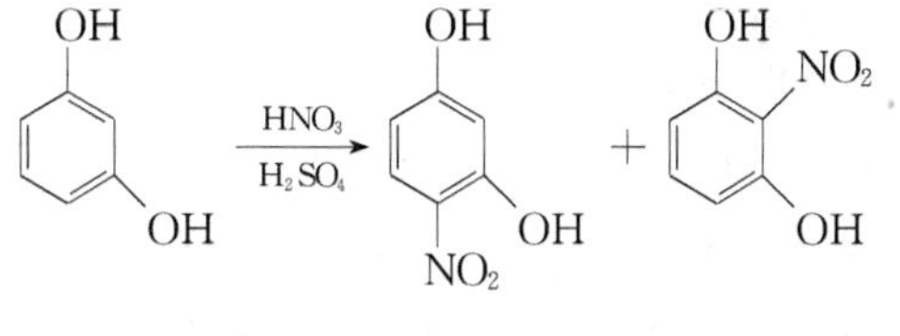

To force the reaction to give 2-nitro-1, 3-benzenediol, or to force the nitration reaction to take place at 2-position, protecting groups can be introduced to 4- and 6-positions prior to the nitration and be removed afterward. Sulfonyl groups is a good candidate to take the 4- and 6-positions via sulfonation and the benzenesulfonic acid can reversibly be converted back to benzene under diluted acidic conditions. By manipulating the reaction conditions, we would achieve the conversion of 1, 3-dihydroxybenzene to 2-nitro-1,3-benzenediol as shown in the following scheme.

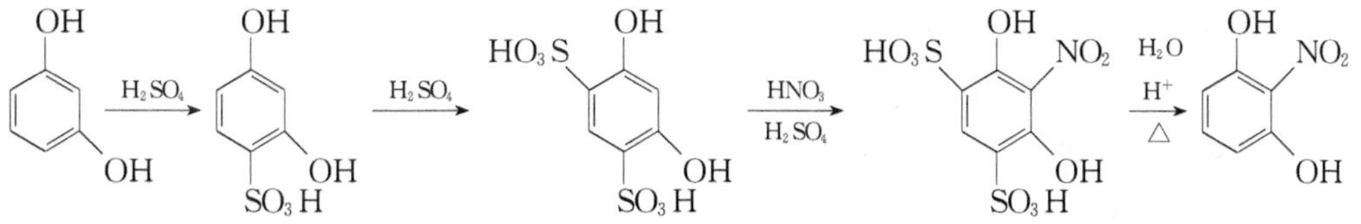

Procedure for synthesis of 2-nitro-1,3-benzenediol①

To 2. 8 g of grounded 1,3-dihydroxybenzene in a 100 mL beaker with a stir bar is slowly and carefully added 13 mL of conc. H_2SO_4 by a Pasteur pipette while being thoroughly stirred. After the addition, the mixture is stirred in a 70 ℃ water bath for additional 15 min. The mixture is then quickly cooled to room temperature using an ice-water bath. The pre-cooled mixture of 2. 8 mL of conc. H_2SO_4 and 2. 0 mL of conc. HNO_3 is dropwise added to the reaction mixture while being stirred and the reaction temperature remains below 25 ℃ with an water bath at ~25 ℃. When the addition completes, the mixture should be stirred in the water bath for another 15 min. Then the mixture is carefully added with ~7 g of crushed ice and the resulting solution is transferred to a two-neck flask in a water bath of ~30 ℃ (better use 1~2 mL of water to rinse the 100 mL beaker and then transfer the water to the two-neck flask carefully). After ~0. 1 g of urea is added to consume the excessive nitric acid, the reaction mixture is subject to steam distillation with the separatory

① The procedure is modified from Robert E. Schaffrath, The Synthesis of 2-Nitroresorcinol, *Journal of Chemical Education*, 1970, 47(3), 224 - 225.

funnel filled with 50 mL of water till there is no further oily distillate collected. When the steam distillation finishes, the distillate is cooled in an ice-water bath and then vacuum filtered to afford the crude product. The crude is transferred to a 25 mL Erlenmeyer flask and then recrystallized using 50% aqueous ethanol solution (~5 mL) to provide the final product. The product is collected by suction filtration, air-dried, and weighed. Dissolve ~1 mg of the product and ~1 mg of 1,3-dihydroxybenzene in ~0.2 mL of ethanol respectively and TLC analysis should be carried out using a running solvent of 20% ethyl acetate in petroleum ether. After the TLC development, the plate should be visualized under UV light at 254 nm. The melting point of the product should also be measured.

Submit the product to the instructor. Clean-up and check-out before leaving the lab.

Technique Checklist

1. Reaction in a hot water bath and then ice-water bath.
2. Magnetic stirring.
3. Steam distillation.
4. Suction filtration.
5. Recrystallization.
6. TLC analysis.
7. Measurement of melting point.

Pre-lab Discussion

1. Mechanism of the electrophilic aromatic substitution and desulfonation.
2. Theory of steam distillation.

Experiment Outline

1. Stir the reaction thoroughly at different temperatures.
2. Perform steam distillation.
3. Perform suction filtration, recrystallization, TLC analysis, and melting point measurement.

Helpful Hints

1. 1,3-Dihydroxybenzene should be grounded in a mortar with a pestle.
2. The sulfonation requires thorough mixing of the solid and the sulfuric acid.
3. The rate to add water to the distillation flask should be similar to the collection rate for the distillate during the steam distillation.
4. The condensing water should be adjusted to flow slowly to avoid the

solidification of the product to block the condenser.

5. The melting point of 2-nitro-1,3-benzenediol is 87～88 ℃.

Results

1. Record the phenomena during your operation.
2. Record the product mass and the TLC result.
3. Record the melting temperature of the product.
4. Calculate the isolated yield of the reaction and the R_f for the product.

5.19 Preparation of the Urea-Formaldehyde Resin

Objectives

1. To prepare urea-formaldehyde resin by a condensation reaction.
2. To master the usage of mechanical stirrer.

Hazards

1. Chemical hazards

Formaldehyde solution (～37%), concentrated ammonia aqueous solution, urea, 1% sodium hydroxide solution, ammonium chloride.

2. Physical hazards

Heating might cause skin-burning and broken glassware may lead to skin damage.

Equipment

1. Condenser.
2. 10 mL and 50 mL Graduated cylinders.
3. Heating set-up.
4. Iron stand and clamps.
5. Mechanical (overhead) stirrer with push-through agitator shafts.
6. Pasteur pipette.
7. Rubber tubing.
8. Stirring rod.
9. Thermometer and adapter.
10. 250 mL Three-neck round-bottom flask.
11. Two wood blocks with clean and smooth surface.

Pre-lab work

Before entering the wet lab, read thoroughly about the experiment and understand the physicochemical principle behind. Plan your set-up of equipment by visiting useful sources and complete the pre-lab report.

Introduction

Polymers are important materials in the modern world and play significant roles in various fields. There are many important polymerization protocols to prepare polymers, such as radical polymerization, olefin ring opening metathesis polymerization, and anionic polymerization, *etc.* One of the widely used polymers is the urea-formaldehyde resin, also referred as urea-methanal resin, which is used in adhesives, molded objects, and particleboard. It belongs to the family of thermosetting resins. This resin material has many favorable properties, such as low water absorption, high tensile strength, flexural modulus, a high heat distortion temperature, high surface hardness, and volume resistance.

The resin is typically synthesized by urea and formaldehyde via a condensation reaction. Under various polymerization conditions, the polymer length and branches can be manipulated. In general, the reaction can be demonstrated below. Simply, two steps are involved: The nucleophilic addition of urea onto the formaldehyde carbonyl group and then dehydration of the intermediate followed by nucleophilic addition of another intermediate to finally provide the polymer. This will lead to a linear polymer chain as shown below.

$$n\,H_2N-\underset{\substack{\|\\O}}{C}-NH_2 + n\,H-\underset{\substack{\|\\O}}{C}-H \longrightarrow n\,H_2N-\underset{\substack{\|\\O}}{C}-NH-CH_2-OH \xrightarrow{-(n-1)H_2O} H\left[-NH-\underset{\substack{\|\\O}}{C}-NH-CH_2-\right]_n OH$$

With additional formaldehyde, the imide group would further react with formaldehyde and lead to crosslink between the polymer chains and this would produce resin of low crosslinking degree.

$$\begin{matrix} \sim NH-\underset{\substack{\|\\O}}{C}-NH\sim \\ \sim NH-\underset{\substack{\|\\O}}{C}-NH\sim \end{matrix} + HCHO \longrightarrow \begin{matrix} \sim NH-\underset{\substack{\|\\O}}{C}-N\sim \\ \quad\quad\quad | \\ \quad\quad\quad CH_2 \\ \quad\quad\quad | \\ \sim NH-\underset{\substack{\|\\O}}{C}-N\sim \end{matrix}$$

Such linear urea-formaldehyde resins of low molecular weight or low crosslinked urea-formaldehyde resins with low molecular weight would be further cured for more complicated network by heating or in presence of curing agents. The cured polymers are usually of high strength and surface hardness, *etc.* The residual hydroxymethyl groups in the polymer contribute to the high adhesive capability of the resin to function as glue.

Procedure for the polymerization

First, 35 mL of formaldehyde solution (~37%) is added to a 250 mL three-neck round-bottom flask equipped with a condenser, a thermometer, and a mechanical stirrer over a water bath. While being stirred, the solution is added with ~ 1.8 mL of concentrated aqueous ammonia solution to adjust the pH to 7.5~8. Then ~11.4 g of urea is slowly added to the mixture. After all urea dissolves, the reaction mixture is slowly heated to 60 ℃ and the reaction remains at this temperature for 15 min, after which it is heated to 97~98 ℃. Then ~ 0.6 g of urea is added to the reaction mixture and the reaction remains at this temperature for ~50 min, during which time the pH of the mixture should be kept at 5.5~6. The heating bath is removed and a small quantity of cold water is added to the mixture. The reaction is cooled below 50 ℃ and ~ 5 mL of the mixture is removed for an adhesion test. The residue of the reaction mixture is added with 1% NaOH solution to adjust the pH to 7~8 and submitted as is.

Adhesion test

To 5 mL of the mixture is added with 0.2 g ammonium chloride and mixed well. Then the mixture is evenly spread on the clean surfaces of two small wood blocks and the wood blocks are pressed together with the gel in the middle. The pair of blocks are labeled appropriately and stored overnight. The adhesion of the resin is checked the second day.

Clean-up and check-out before leaving the lab.

Technique Checklist

1. Mechanical stirring.
2. Addition of reagents to the reaction.

Pre-lab Discussion

1. Theory of the polymerization reaction.
2. Key points of using mechanical stirrer.

Experiment Outline

1. Use mechanical stirring to thoroughly mix the chemicals, even with increasing viscosity.
2. Monitor the pH of the reaction mixture constantly.
3. Add urea to the reaction mixture in portions.
4. Test the adhesion property of the polymer.

Helpful Hints

1. It is important to keep the reaction temperature stable during the reaction. Abrupt temperature dropping may lead to drastic viscosity increase.

2. To achieve minimum temperature change of the reaction mixture, urea should be added slowly.

3. The pH of the reaction mixture should be constantly checked and adjusted.

4. In order to determine the reaction endpoint, some of the reaction mixture may be removed to examine the viscosity.

5. If viscosity increases abruptly, the reaction should be cooled down immediately, the reaction solution might be diluted with some aqueous formaldehyde solution, or NaOH might be added to adjust the pH of the mixture to ~ 7.

Results

1. Record the reaction phenomena.
2. Describe the adhesion property of the polymer.
3. Describe the final mixture that is submitted.

5.20 Preparation of *Cyclo*hexene

Objectives

1. To understand the mechanism of intramolecular dehydration of alcohol.
2. To practice fractional distillation.

Hazards

1. Chemical hazards

*Cyclo*hexanol, conc. phosphoric acid, sodium chloride, 5% sodium carbonate solution, brine, anhydrous calcium chloride, boiling chips, 1% bromine in CCl_4.

2. Physical hazards

Heating might cause skin-burning and broken glassware may lead to skin damage.

Equipment

1. Distillation head.
2. 50 mL Erlenmeyer flask.
3. 10 mL Graduate cylinder.
4. Heating set-up.

5. Ice-water bath.
6. Iron ring, iron stand and clamps.
7. Liebig Condenser.
8. Pasteur pipette.
9. Receiver adapter.
10. 50 mL Round bottom flask.
11. Rubber tubing.
12. 125 mL Separatory funnel.
13. Thermometer with adapter.
14. Vigreux distillation column.

Pre-lab work

Before entering the wet lab, read thoroughly about the experiment and understand the physicochemical principle behind. Plan your set-up of equipment by visiting useful sources and complete the pre-lab report.

Introduction

Olefins, containing carbon-carbon double bonds, are regarded as important synthetic building blocks and demonstrate significant applications in various scientific fields. There are various ways to achieve olefin production, one of which is from the elimination of alcohols or alkyl halides. The elimination of alcohols is usually carried out under acidic conditions and the reaction temperature must be carefully controlled to avoid the competitive intermolecular dehydration. Sulfuric acid or phosphoric acid is usually used to catalyze the reaction with high efficiency, which can provide protons to the reaction as well as effectively remove water from the intermediate. For instance, *cyclo*hexanol can be treated with conc. sulfuric acid or phosphoric acid to prepare *cyclo*hexene with a moderate yield. A fractional distillation can be applied to efficiently remove the products from the reaction mixture.

$$\text{cyclohexanol (OH)} \xrightarrow[\Delta]{H_3PO_4} \text{cyclohexene} + H_2O$$

Procedure for the reduction

In a 50 mL round bottom flask are placed 10 g of *cyclo*hexanol, 4 mL of conc. phosphoric acid, and several boiling chips. The mixture is thoroughly mixed. The round bottom flask is attached with a Vigreux distillation column, a distillation head, a thermometer, a condenser, a receiver adapter, and a receiving Erlenmeyer flask for fractional distillation. The receiving Erlenmeyer flask is immersed in an ice-water bath and the reaction flask is heated to reflux during which the temperature at the distillation head should remain no higher than 90 ℃. The distillation is terminated when the temperature of distillate increases or nearly no further distillate at 90 ℃ is collected. The distillate is added with 1 g of sodium chloride and mixed thoroughly. Then, 4 mL of 5% sodium carbonate solution is added to neutralize the mixture. The mixture is transferred to the separatory funnel for layer separation. The organic layer is further washed with 5 mL of water and 5 mL of brine, and then dried over anhydrous calcium chloride. The solid is filtered off and the filtrate is subject to atmospheric distillation for further purification.

Characteristic test with bromine

Either 0. 2 mL of the product or 0. 2 mL of *cyclo*hexanol is added with 1 drop of 1% bromine in carbon tetrachloride to identify the reactivity of the compounds. ①

Submit the product to the instructor. Clean-up and check-out before leaving the lab.

Technique Checklist

1. Fractional distillation to remove the product from the reaction mixture.
2. Extraction.
3. Atmospheric distillation.
4. Drying a solution with anhydrous reagent.

Pre-lab Discussion

1. The mechanism of elimination of *cyclo*hexanol.
2. The principle of fractional distillation.

Experiment Outline

1. Reaction proceeds along with fractional distillation.
2. Neutralization with sodium carbonate solution.
3. Separate and wash the organic layer with water and brine respectively using a separatory funnel.
4. Perform atmospheric distillation to collect the desired product.
5. Add bromine to the product/precursor.

Helpful Hints

1. The heating source should be adjusted to assist the fractional distillation.

---

① This test depends on the availability of bromine in the undergraduate laboratory. As bromine is volatile, this test must be undertaken in a hood and the source bottle must be capped tightly after use.

2. Insulation of the Vigreux distillation column with cotton or aluminium foil can accelerate the fractional distillation.

3. The boiling point of *cyclo*hexene is 82～83 ℃.

4. All glassware for the reaction and atmospheric distillation should be dry.

5. *Cyclo*hexene and water can form azeotrope with a weight percentage of 58.4% with a boiling point at 70.8 ℃. *Cyclo*hexene and *cyclo*hexanol can form azeotrope with a boiling point at 64.9 ℃. *Cyclo*hexanol and water can form azeotrope with a boiling point at 97.8 ℃.

Results

1. Record the reaction phenomena.
2. Record the mass and boiling point for the product.
3. Record the phenomenon for the characteristic test.
4. Calculate the isolated yield of the reaction.

5.21 Synthesis of 2-Chloro-2-methylpropane

Objectives

1. To prepare alkyl halide using the S_N1 reaction.
2. To practice extraction and distillation.

Hazards

1. Chemical hazards

Conc. HCl, *tert*-butanol, 5% sodium bicarbonate aqueous solution, sodium bicarbonate, anhydrous calcium chloride.

2. Physical hazards

Heating might cause skin-burning and broken glassware may lead to skin damage.

Equipment

1. 500 mL Beaker.
2. Distillation head and receiver adapter.
3. 50 mL Erlenmeyer flask.
4. 10 mL Graduated cylinder.
5. Heating set-up.
6. Iron ring, iron stand and clamps.
7. Liebig condenser.

8. Magnetic stirrer with a suitable magnetic stir bar.
9. 50 mL Round bottom flask.
10. Reflux condenser.
11. Rubber tubing.
12. 125 mL Separatory funnel.
13. Thermometer with thermometer adapter.

Pre-lab work

Before entering the wet lab, read thoroughly about the experiment and understand the physicochemical principle behind. Plan your set-up of equipment by visiting useful sources and complete the pre-lab report.

Introduction

Alkyl halides are useful alkylation reagents, which are widely used in organic synthesis. There are several methods to prepare alkyl halides, such as nucleophilic substitution reactions, radical substitution reaction, and electrophilic addition reactions, *etc.*, among which nucleophilic substitution reactions are one of the most used. The frequently used nucleophilic substitution reaction for synthesizing alkyl halides is by treating alcohols with concentrated aqueous solutions of hydrogen halide. Either monomolecular or bimolecular nucleophilic substitution reaction can be carried out depending on the type of alcohols. Typically, *tertiary* alcohols prefers the S_N1 reaction. For instance, *tert*-butyl chloride can be prepared via *tert*-butanol as shown below.

$$(CH_3)_3C\text{-}OH + HCl \longrightarrow (CH_3)_3C\text{-}Cl + H_2O$$

Procedure for the substitution

Add 25 mL of conc. HCl to 10 mL of *tert*-butanol in a 50 mL round bottom flask placed with a stir bar. A condenser is attached to the round bottom flask and the mixture is stirred vigorously for 50 min at room temperature. Stop the stirring to observe if two layers form. If two layers are present, the mixture is transferred to a separatory funnel and pour the aqueous layer into a 500 mL beaker as a waste, which should be neutralized at the end of the experiment before disposal①. The organic layer is then washed with 10 mL of fresh water in the separatory funnel. The organic layer

① If not, keep the reaction stirring for another 30 min and then check.

is separated and washed with 5% sodium bicarbonate followed by 10 mL of water. ① The organic layer is collected in a dry 50 mL Erlenmeyer flask and dried over anhydrous $CaCl_2$. The solid is removed by decanting the liquid into a dry round-bottom flask and the liquid is subject to atmospheric distillation to collect the final product, during which the receiving flask should be cooled over an ice-water bath to avoid loss of the product to atmosphere.

Technique Checklist

1. Magnetic stirring.
2. Extraction.
3. Distillation.

Pre-lab Discussion

Mechanism of the monomolecular nucleophilic substitution.

Experiment Outline

1. Use magnetic stirring to mix thoroughly the reactants.
2. Separate and washing the organic layer by extraction.
3. Dry the organic layer with anhydrous $CaCl_2$.
4. Carry out atmospheric distillation.

Helpful Hints

1. The boiling point of 2-chloro-2-methylpropane is 50～51 ℃.
2. Calcium chloride is known to form complex with alcohol.

Results

1. Record the reaction phenomena.
2. Record the mass and boiling point for the product.
3. Calculate the isolated yield of the reaction.

5.22 Synthesis of Dibutyl Ether

Objectives

1. To synthesize ether via the intermolecular dehydration.

---

① All aqueous layers are combined carefully in the 500 mL beaker and then further neutralized with sodium bicarbonate powder to pH ～7 before it is disposed to the sink.

2. To enhance the reaction by azeotropic removal of H_2O using the Dean-Stark apparatus.

3. To practice the simple distillation and extraction procedures.

Hazards

1. Chemical hazards

Boiling chips, butan-1-ol, conc. H_2SO_4, 50% H_2SO_4 aqueous solution, anhydrous calcium chloride.

2. Physical hazards

Broken glassware may lead to skin damage and heating might cause skin-burning.

Equipment

1. Boiling chips.
2. Dean-Stark apparatus.
3. Distillation head and receiver adapter.
4. 100 mL or 50 mL Erlenmeyer flasks.
5. 10 mL or 25 mL Graduated cylinder.
6. Heating set-up.
7. Ice-water bath.
8. Iron ring, iron stand and clamps.
9. Liebig condenser.
10. Oil bath.
11. 50 mL Round bottom flask.
12. Rubber tubing.
13. 125 mL Separatory funnel.
14. Thermometer and adapter.
15. 50 mL Three-neck flask.

Pre-lab work

Before entering the wet lab, read thoroughly about the experiment and understand the physicochemical principle behind. Plan your set-up of equipment by visiting useful sources and complete your pre-lab report.

Introduction

Alcohols can undergo intermolecular dehydration reactions to provide ethers. The reaction is undertaken under acidic conditions, which facilitates the installation of a good leaving group: Protonation of hydroxyl group can yield a good leaving group as water. On the other hand, the hydroxy group in alcohols is nucleophilic and can attack the newly formed electrophilic protonated alcohol molecule immediately once it forms. Such intermolecular dehydration can be either S_N1 or S_N2 depending on the structures of alcohols. This method is useful to prepare symmetric alcohols. The intramolecular dehydration is the common competing reaction along the reaction process, so that the reaction temperature should be carefully controlled to suppress side reactions. For instance, dibutyl ether can be formed from butan-1-ol under 135 ℃ while but-1-ene is the major product at higher temperatures.

$$\xleftarrow[>135\ ^\circ C]{H_2SO_4} \quad OH \quad \xrightarrow[134\sim135\ ^\circ C]{H_2SO_4} \quad O$$

Procedure

A 50 mL three-neck flask with some boiling chips is equipped with a thermometer and a Dean Stark apparatus that is filled with water up to 90% of its side-arm volume. ① And 16 mL of butan-1-ol and 2.5 mL of conc. H_2SO_4 are added to the flask and a stopper is used to cap the flask. On the other hand, a condenser is attached to the Dean Stark apparatus with running water flowing. The reaction mixture is heated to ~ 135 ℃ in an oil bath for approximately 1 hour until ~ 2 mL of additional water is collected at the side arm of the Dean Stark apparatus. The reaction mixture is cooled to room temperature and transferred to a separatory funnel containing 25 mL of water. Being shaken thoroughly, the separatory funnel is left still to allow the layers to separate. The organic layer is then further washed with 15 mL of 50% H_2SO_4 solution twice and 10 mL of water, and dried over anhydrous calcium chloride. The organic crude is filtered and subject to atmospheric distillation to collect the desired product.

Submit the product to the instructor. Clean-up and check-out before leaving the lab.

Technique Checklist

1. Removal of water azeotropically.
2. Extraction.
3. Distillation.

Pre-lab Discussion

1. Mechanism of the synthesis of dibutyl ether.
2. Possible side-products and cause.

① (V-2) mL of water in the side-arm of the Dean-Stark apparatus is usually good where V is the volume of the side-arm.

Experiment Outline

1. Heat the reaction with a Dean-stark apparatus attached.
2. Perform atmospheric pressure distillation and extraction.
3. Weigh the crude product and purified product.
4. Record the boiling point of the product.

Helpful Hints

1. Make sure all joints are sealed well. Otherwise, one will lose the product into the atmosphere.

2. Do not overheat the mixture, or one will experience a poor conversion yield and severe side reactions.

3. The boiling point of dibutyl ether is 140～141 ℃.

4. The azeotropic boiling point of butan-1-ol and water is 92. 7 ℃ and that of dibutyl ether and water is 92. 9 ℃.

Results

1. Record the boiling temperature and amount of the crude and purified liquids.
2. Record the phenomena accompanying the procedure.
3. Calculate the isolated yield of the procedure.

5.23 Synthesis of Butyl Phenyl Ether

Objectives

1. To synthesize an ether via the Williamson ether synthesis.
2. To apply phase transfer catalysis to facilitate a reaction.
3. To apply mechanic stirring for a reaction while refluxing.
4. To practice extraction procedure to remove inorganic components.
5. To practice simple distillation and switch condensers during the course.

Hazards

1. Chemical hazards

Phenol, sodium hydroxide, 1-bromobutane, 6% tetrabutylammonium bromide, 10% sodium hydroxide solution, brine, anhydrous sodium sulfate, water, ethyl acetate, petroleum ether, dichloromethane.

2. Physical hazards

Broken glassware may lead to skin damage and heating might cause skin-burning.

Equipment

1. Air condenser.
2. Dean-Stark apparatus.
3. Distillation head and receiver adapter.
4. 100 mL or 50 mL Erlenmeyer flasks.
5. 25 mL Graduated cylinder.
6. Heating set-up.
7. Iron ring, iron stand and clamps.
8. Liebig condenser.
9. Mechanical (overhead) stirrer with push-through agitator shafts.
10. 50 mL Round bottom flask.
11. Rubber tubing.
12. 125 mL Separatory funnel.
13. Test tubes.
14. Thermometer and adapter.
15. 100 mL Three-neck round-bottom flask.
16. TLC chamber and alumina foil.
17. TLC plate and capillary tubes.
18. UV light at 254 nm.

Pre-lab work

Before entering the wet lab, read thoroughly about the experiment and understand the physicochemical principle behind. Plan your set-up of equipment by visiting useful sources and complete your pre-lab report.

Introduction

In addition to the intermolecular dehydration of alcohol, ethers can also be prepared based on the Williamson reaction, which uses alkoxide (or phenoxide) as a nucleophile to attack alkyl halide to form the ether bond. The Williamson ether synthesis is powerful to make a mixed ether. The Williamson reaction can be used to prepare alkyl phenyl ethers but the intermolecular dehydration method is powerless for such target molecules. However, the alkoxide or phenoxide generally has low solubility in the organic phase because of its ionic nature and its actual concentration in the organic phase however would determine the reaction rate since the substitution reaction for ether production is bimolecular. Efforts have been made to improve the solubility of such salts in organic phase, such as using different inert organic solvents, increasing the reaction temperatures, and/or involving phase transfer agents to form ionic pairs with the alkoxide or phenoxide and help it transport into the organic phase. Because the phase transfer agent is inert and it will be released and go back to the aqueous phase when the alkoxide or phenoxide is consumed in the organic layer, the phase transfer agent can participate in cycles to transport the alkoxide or phenoxide from the aqueous phase into the organic phase. This is usually referred as *phase transfer catalysis*. An illustrative scheme for the phase transfer catalysis to prepare an alkyl phenyl ether is demonstrated below.

$PhONa + Q^+Br^- \rightleftharpoons PhO^-Q^+ + NaBr$ aq. phase

$PhOC_4H_9 + Q^+Br^- \rightleftharpoons PhO^-Q^+ + C_4H_9Br$ org. phase

Procedure

To a 100 mL three-neck round-bottom flask equipped with a mechanic stirrer and a condenser are added 6.3 g of phenol, 5.2 g of sodium hydroxide, 5 mL of 6% tetrabutylammonium bromide, and 25 mL of water. The mixture is stirred at room temperature for 5 min until all solid dissolves. To the mixture, 15.0 mL of 1-bromobutyl is added in one portion and a stopper is used to cap the three-neck round-bottom flask. The flask is heated to reflux for 70 min while the mechanical stirrer works vigorously. Then the stirring and heating is turned off. Upon cooling, the mixture is transferred to a separatory funnel and the organic phase is collected and washed with 2×10 mL of 10% sodium hydroxide solution and 3×10 mL of brine separately. The organic layer is separated and dried over anhydrous sodium sulfate for 10 min. The solid is filtered off and the filtrate is subject to simple distillation with a Liebig condenser to collect the fraction with boiling points lower than 140 ℃. Then the Liebig condenser is replaced by an air condenser and the distillation flask is heated to collect the fraction with a boiling point of 200 ~ 210 ℃ in a pre-weighed Erlenmeyer flask. The fraction with the higher boiling point is weighed.

TLC analysis

Dissolve 5 mg of phenol and/or the product in 1 mL of dichloromethane and loaded onto a TLC plate. After development under 1 : 3 ethyl acetate: petroleum ether in the TLC chamber, the TLC plate is visualized under the UV light at 254 nm.

Submit the product to the instructor. Clean-up and check-out before leaving the lab.

Technique Checklist

1. Reaction at reflux with a mechanic stirrer.
2. Extraction.
3. Distillation using a Liebig condenser and then an air condenser.
4. TLC analysis.

Pre-lab Discussion

1. Mechanism of the synthesis of butyl phenyl ether.
2. Principle of phase transfer catalysis.

Experiment Outline

1. Assemble the glassware for the reaction.
2. Mechanically stir and heat the reaction at ~ 70 ℃ for 70 min.
3. Wash the reaction mixture with NaOH solution and brine separately.
4. Perform atmospheric pressure distillation.
5. Perform TLC analysis.

Helpful Hints

1. Make sure all joints are sealed well. Otherwise, one will lose the product into the atmosphere.
2. The boiling point of butyl phenyl ether is 210 ℃.

Results

1. Record the boiling temperature and mass of the purified liquid.
2. Record the phenomena accompanying the procedure and the TLC result.
3. Calculate the isolated yield of the procedure and R_f of the product.

5.24 Synthesis of Hexane-1,6-dionic Acid

Objectives

1. To use potassium permanganate as an oxidizing reagent.
2. To prepare hexane-1,6-dionic acid, also called adipic acid, from *cyclo*hexene.

Hazards

1. Chemical hazards

*Cyclo*hexene, potassium permanganate, water, concentrated hydrochloric acid,

1% sodium hydroxide solution.

2. Physical hazards

Heating might cause skin-burning and broken glassware may lead to skin damage.

Equipment

1. Büchner funnel and suction flask.
2. 10 mL and 50 mL Graduated cylinder.
3. 250 mL Beaker.
4. 250 mL Erlenmeyer flask with a magnetic stir bar.
5. Filter paper and pH paper.
6. Heating set-up.
7. Ice-water bath.
8. Iron stand and clamps.
9. Magnetic stirrer.
10. Melting point apparatus and melting tube.
11. Reflux condenser.
12. Rubber adapter.
13. Stirring rod.
14. Water bath.
15. Water pump.

Pre-lab work

Before entering the wet lab, read thoroughly about the experiment and understand the physicochemical principle behind. Plan your set-up of equipment by visiting useful sources and complete your pre-lab report.

Introduction

Potassium permanganate is a widely used strong oxidizing reagent. It is powerful to oxidize olefins, alcohols, and aldehydes, *etc*. Using potassium permanganate to oxidize olefins or aldehydes is a classic method to prepare carboxylic acids. Examples for the cleavage of olefins by potassium permanganate are shown in the following equations. It is clear that permanganate oxidation can afford carboxylic acids, ketones, or carbon dioxide depending on the olefin's structure.

$$\xrightarrow[H_2O]{KMnO_4} CH_3COOH + CO_2 \qquad \xrightarrow[H_2O]{KMnO_4} CH_3COOH +$$

When a *cyclo*alkene is the substrate, permanganate can also oxidize the double bond but the two carbonyl groups (either ketone carbonyl or carboxyl group) are present at the termini of the same molecule. Therefore, oxidization of *cyclo*alkene is a nice method to introduce two functional groups to the molecule at meanwhile. For instance, *cyclo*hexene can be oxidized to give hexane-1,6-dioic acid.

$$\text{cyclohexene} \xrightarrow[H_2O]{KMnO_4} HOOC(CH_2)_4COOH$$

Procedure

To a 250 mL Erlenmeyer flask are added 2 mL of *cyclo*hexene, 8. 4 g of potassium permanganate, and 50 mL of water. The mixture is stirred vigorously over a 45 ℃ water bath for 30 min. The solution is then heated to reflux for 15 min. Then portions of 1 mL of methanol are added to the reaction mixture until the reaction solution is colorless. ① The reaction is cooled to room temperature and filtered under vacuum. Subsequently, 2×10 mL of hot 1% sodium hydroxide solution is added to rinse the Erlenmeyer flask and poured to wash the filter cake. All filtrate is combined in a 250 mL beaker and boiled while being stirred using a stir rod until the solution volume is approximately 10 mL. The resulting solution is cooled in an ice-water bath and acidified to pH ~1 carefully and slowly using conc. hydrochloric acid while thoroughly stirring the solution using a stir rod. The beaker is allowed to stand in the ice-water bath for 10 min for the product to completely crystallize. The solid is collected by vacuum filtration and recrystallized using no more than 5~10 mL of boiling water.

The melting point of the product should be measured to further characterize the product.

Submit the product to the instructor. Clean-up and check-out before leaving the lab.

Technique Checklist

1. Reaction at high temperatures.
2. Precipitation by pH adjustment.
3. Vacuum filtration and recrystallization.
4. Melting point measurement.

Pre-lab Discussion

1. Conditions of the permanganate oxidation.

---

① There may also be particles suspending in the solution and one should try to observe the color change of the solution carefully.

2. Role of methanol added into the reaction mixture.

Experiment Outline

1. Mix and warm *cyclo*hexene, potassium permanganate and water.
2. Heat the reaction to reflux further.
3. Filter the reaction mixture to remove the side product manganese dioxide.
4. Acidify the reaction mixture to precipitate the product.
5. Recrystallize the product using water.
6. Measure the melting point.

Helpful Hints

1. To check if the reaction is colorless, one can use a stir rod to tip a bit of the reaction mixture onto a filter paper and check if a purple ring or colorless ring appears around the dark brown spot of manganese dioxide.

2. Methanol is added to destroy the excess of potassium permanganate. Typically, if there is permanganate present, 1 mL of methanol is added and the mixture should be heated. This may be repeated until all permanganate is consumed.

3. Addition of conc. hydrochloric acid should be carried out in the fumehood.

4. The melting point of hexane-1,6-dioic acid (adipic acid) is 153 ℃.

Results

1. Record the phenomena during your operation.
2. Record the amount of the crude product and recrystallized product.
3. Record the melting point of the product.
4. Calculate the isolated yield of the reaction.

5.25 Preparation of Furfuryl Alcohol and 2-Furoic Acid by the Cannizzaro Reaction

Objectives

1. To understand the mechanism of the Cannizzaro reaction.
2. To practice atmospheric distillation and recrystallization.
3. To characterize product using thin layer chromatography.

Hazards

1. Chemical hazards

43% sodium hydroxide, furan-2-carbaldehyde, diethyl ether, anhydrous magnesium sulfate, conc. hydrochloric acid, 1 : 1 EtOAc: petroleum ether, 1 : 9 methanol: dichloromethane, water.

2. Physical hazards

Heating might cause skin-burning and broken glassware may lead to skin damage.

Equipment

1. 50 mL Beaker.
2. Büchner funnel and suction flask.
3. Distillation head, Liebig condenser, and receiver adapter.
4. 50 mL Erlenmeyer flasks.
5. Filter paper and pH paper.
6. Glass rod.
7. 25 mL Graduate cylinder.
8. Heating set-up.
9. Ice-water bath.
10. Iron ring, iron stand and clamps.
11. Magnetic stirrer and a suitable magnetic stir bar.
12. Melting point apparatus and melting tubes.
13. Pasteur pipette.
14. 100 mL Round bottom flask.
15. Rubber adapter.
16. 125 mL Separatory funnel.
17. Test tubes.
18. Thermometer with an adapter.
19. TLC plate, TLC chamber, aluminium foil and capillary tube.
20. UV light at 254 nm.
21. Water pump.

Pre-lab work

Before entering the wet lab, read thoroughly about the experiment and understand the physicochemical principle behind. Plan your set-up of equipment by visiting useful sources and complete the pre-lab report.

Introduction

The Cannizzaro reaction is a useful base-catalyzed redox disproportionation of an aldehyde having no α-hydrogen atoms. The reaction provides the oxidized product as a carboxylic acid and the reduced product as an alcohol. Sodium hydroxide and potassium hydroxide are the frequently used strong bases and the reaction can take place in the presence of water. The reaction scheme studied in this experiment is demonstrated below.

$$2\,\text{(furan-2-yl)}-CHO \xrightarrow[2.\ H_3O^+]{1.\ NaOH} \text{(furan-2-yl)}-CH_2OH + \text{(furan-2-yl)}-COOH$$

Procedure

A 50 mL Erlenmeyer flask containing 6.6 mL of freshly distilled furanyl-2-carbaldehyde is dropwise added with 6 mL of pre-cooled 43% sodium hydroxide solution over an ice-water bath for about 10 min. The resulting mixture is stirred over an ice-water bath for 20 min and 10 mL of water is added to the reaction mixture. The aqueous solution is transferred to a separatory funnel and extracted with 15 mL of fresh diethyl ether three times. The ether layers are combined and dried over anhydrous magnesium sulfate while the aqueous layer is collected in a 50 mL beaker.

For the ether layer, after the removal of magnesium sulfate by filtration, the filtrate is concentrated by an atmospheric distillation and then the alcohol is collected at ~ 170 ℃.

Conc. hydrochloric acid is added carefully and slowly to acidify the aqueous layer from the extraction to pH 2 ~ 3. The solid is collected by vacuum filtration and washed with a small quantity of water. The crude solid is then subject to recrystallization using water and the final product is collected by vacuum filtration and dried over air.

The products should be characterized by boiling point or melting point measurement and TLC analysis. For TLC analysis, dissolve ~2 mg of the crystal in 1 mL dichloromethane and load it onto a TLC plate. Then the plate is developed using 1 : 1 EtOAc: petroleum ether and/or 1 : 9 methanol: dichloromethane as the developing solvent. After TLC development, the plate should be visualized under UV light at 254 nm.

Infrared, ultraviolet-visible and NMR spectroscopic analysis is suggested to characterize the products.

Submit the products to the instructor. Clean-up and check-out before leaving the lab.

Technique Checklist

1. Magnetic stirring.
2. Extraction, atmospheric distillation, recrystallization, and suction filtration.

3. Melting point measurement and TLC analysis.

Pre-lab Discussion

Mechanism of the Cannizzaro reaction.

Experiment Outline

1. Slowly add cold sodium hydroxide solution to the aldehyde.
2. Undertake the reaction over an ice-water bath.
3. Perform extraction of the alcohol using diethyl ether and then purify the alcohol using atmospheric distillation.
4. Acidify the aqueous layer using hydrochloric acid and collect the acid by vacuum filtration.
5. Recrystallize the solid crude with water.
6. Analyze the purity of the products by TLC and melting point test.

Helpful Hints

1. The reaction mixture should be cooled prior to the addition of sodium hydroxide.
2. The boiling point of furfuryl alcohol is 170～171 ℃.
3. The melting point of 2-furoic acid is 129～130 ℃.

Results

1. Record the reaction phenomena.
2. Record the mass, melting point, and TLC result for the products.
3. Calculate the isolated yield of the reaction and the R_f's of the products.

5.26 A Solvent Free Cannizzaro Reaction

Objectives

1. To understand the mechanism of the Cannizzaro reaction.
2. To practice the Cannizzaro reaction under solvent free conditions.
3. To pestle the mixture to accelerate the reaction.
4. To characterize product using thin layer chromatography.

Hazards

1. Chemical hazards

Potassium hydroxide, 2-chlorobenzaldehyde, ethanol, conc. hydrochloric acid,

50% ethyl acetate in petroleum ether, 5% methanol in dichloromethane, water

2. Physical hazards

Heating might cause skin-burning and broken glassware may lead to skin damage.

Equipment

1. 50 mL Beaker.
2. Büchner funnel and suction flask.
3. Filter paper and pH paper.
4. Glass rod.
5. 10 mL Graduate cylinder.
6. Iron stand and clamps.
7. Magnetic stirrer and a suitable magnetic stir bar.
8. Melting point apparatus and melting tubes.
9. Pasteur pipette.
10. Pestle and mortar.
11. Rubber adapter.
12. Test tubes.
13. TLC plate, TLC chamber, aluminium foil and capillary tube.
14. UV light at 254 nm.
15. Water pump.

Pre-lab work

Before entering the wet lab, read thoroughly about the experiment and understand the physicochemical principle behind. Plan your set-up of equipment by visiting useful sources and complete the pre-lab report.

Introduction

The Cannizzaro reaction requires strong base to achieve the redox disproportionation of aldehydes containing no α-hydrogen atoms. The normal reaction conditions usually adopt concentrated aqueous solutions of bases, which may not be reasonable to achieve a good homogeneous reaction conditions as the aldehydes are usually less soluble in aqueous conditions. Accordingly, either water-miscible organic solvent, such as tetrahydrofuran or alcohols, needs to be used to facilitate the reaction, or high temperatures might be involved. On the other hand, it has been found that some reactions can be accelerated along with the load of mechanical forces and the mechanochemistry is somehow very useful in certain reactions to facilitate the cleavage of covalent bonds and formation of novel covalent bonds.

As in this experiment, a mechanically triggered Cannizzaro reaction is undertaken in the presence of potassium hydroxide with *ortho*-chlorobenzaldehyde. Along with the pestling process, the base and the aldehyde are mixed thoroughly to achieve the disproportionation of the aldehyde.

$$2\ \text{2-ClC}_6\text{H}_4\text{CHO} \xrightarrow[\text{2. } H_3O^+]{\text{1. KOH Pestling}} \text{2-ClC}_6\text{H}_4\text{COOH} + \text{2-ClC}_6\text{H}_4\text{CH}_2\text{OH}$$

Procedure

In a mortar is added 2 mL of 2-chlorobenzaldehyde and 1. 5 g of potassium hydroxide. The mixture is then grounded using a pestle continuously. After approximately 30 min, a little of the reaction mixture, which should appear like thick paste, is taken out to suspended in 0. 5 mL of ethanol for TLC analysis. Then the mixture is added with 10 mL water and stirred thoroughly. The suspension is then filtered and the solid is further washed with 5 mL of water twice and dried in the air as product 1. All the filtrate is collected and combined. Concentrated hydrochloric acid is carefully added to acidify the filtrate to pH ~1. The resulting mixture is filtered and washed thoroughly with water and dried in the air as product 2.

Both products should be characterized by melting point measurement and TLC analysis. For TLC analysis, dissolve ~5 mg of the products in 1 mL ethanol separately and load them onto a TLC plate. Then the plate is developed using 50% ethyl acetate in petroleum ether and/or 5% methanol in dichloromethane as the developing solvent. After TLC development, the plate should be visualized under UV light at 254 nm.

Infrared, ultraviolet-visible and/or NMR spectroscopic analysis is suggested to characterize the products and *ortho*-chlorobenzaldehyde.

Submit the products to the instructor. Clean-up and check-out before leaving the lab.

Technique Checklist

1. Operation of grounding using the pestle and mortar.
2. Acidification and suction filtration.
3. Melting point measurement and TLC analysis.

Pre-lab Discussion

Mechanism of the Cannizzaro reaction.

Experiment Outline

1. Ground the mixture of potassium hydroxide and the aldehyde for 30 min.
2. PerformTLC analysis of the reaction mixture.
3. Add water to the reaction .
4. Filter the mixture in vacuum.
5. Acidify the filtrate and suction filter the mixture.
6. Analyze the purity of the product by TLC and melting point test.

Helpful Hints

1. The reaction mixture should change from a gummy mixture into a thick paste.
2. The melting point of *ortho*-chlorobenzyl alcohol is 73. 0 ℃.
3. The melting point of *ortho*-chlorobenzoic acid is 140～142 ℃.

Results

1. Record the reaction phenomena.
2. Record the masses, melting points, and TLC results for the products.
3. Calculate the isolateds yields of the reaction and the R_f's of the products.
4. Identify the structures for product 1 and 2.

5.27 Reduction of 3-Nitroacetophenone Using Sodium Borohydride

Objectives

1. To understand the chemoselectivity of $NaBH_4$.
2. To practice a reduction reaction using $NaBH_4$.
3. To use rotatory evaporation under reduced pressure to remove volatiles.

Hazards

1. Chemical hazards

Sodium borohydride, 3-nitroacetophenone, ethanol, toluene, ethyl acetate, anhydrous sodium sulfate, 1 : 1 ethyl acetate: petroleum ether.

2. Physical hazards

Heating might cause skin-burning and broken glassware may lead to skin damage.

Equipment

1. Büchner funnel and suction flask.

2. Condenser.
3. 25 mL Graduate cylinder.
4. Filter paper and rubber adapter.
5. Heating set-up.
6. Ice-water bath.
7. Iron ring, iron stand and clamps.
8. Magnetic stirrer with a suitable magnetic stir bar.
9. Melting point apparatus and melting tubes.
10. Rota-evaporator with a chilling circulation system.
11. 50 mL Round bottom flask.
12. Separatory funnel.
13. TLC plate, TLC chamber, aluminium foil and capillary tube.
14. UV light at 254 nm.
15. Water pump.

Pre-lab work

Before entering the wet lab, read thoroughly about the experiment and understand the physicochemical principle behind. Plan your set-up of equipment by visiting useful sources and complete the pre-lab report.

Introduction

Reduction is one of the important reactions used in organic synthesis. There are various reduction methods, such as catalytic hydrogenation, reduction via hydride transfer agents, and dissolving metal reduction. All these methods are applied to achieve various reduction purposes: Converting aldehydes/ketones to alcohols, converting esters/amides to aldehydes or alcohols, converting aldehydes/ketones to alkanes, or converting unsaturated carbon-carbon multiple bonds into single bonds, *etc.* Hydride transfer agents are commonly used to reduce a large number of compounds, which containing polar unsaturated bonds, such as carbonyl groups and cyano-groups. The hydride is usually transferred to these unsaturated bonds via a nucleophilic addition mechanism, nevertheless that different hydride transfer agents represent different reactivities and selectivities. Among many hydride transfer agents, lithium aluminium hydride and sodium borohydride are the best known and most used, the former generally more reactive than the latter. Lithium aluminium hydride, of a highly reactivity, must be used in anhydrous aprotic solvent and can reduce nearly all polarized multiple bonds, and its use must be with high precautions. On the other hand, sodium borohydride is less reactive and can be used in various protic solvents like alcohols and even water. As sodium borohydride is a mild reducing agent, it displays a relatively great selectivity on different functional groups. Sodium borohydride can reduce acyl halides, aldehydes, and ketones into corresponding alcohols but it is relatively inert to groups such as nitro, cyano, ester, and amide. As a result, sodium borohydride can be used to reduce the target carbonyl group in aldehyde/acyl halides/ketones in presence of functional groups that are inert under the conditions. One example, which is the key conversion in this experiment, is shown below as the carbonyl group is preferentially reduced over the nitro group under the defined conditions.

NO_2 NO_2

$\xrightarrow[EtOH]{NaBH_4}$

O OH

Procedure for the reduction①

Dissolve 1. 65 g of 3-nitroacetophenone in 20 mL of ethanol in a 50 mL round bottom flask equipped with a stir bar in an ice-water bath. While the solution is stirred in the ice-water bath, 0. 45 g of sodium borohydride is added in small portions over 5 min and the reaction is further stirred for 20 min. The reaction mixture is added with 15 mL of water and heated to boil for 2 min and then cooled to room temperature. The resulting mixture is transferred to a separatory funnel and extracted with 2×30 mL of ethyl acetate. The organic layers are combined, further washed with 30 mL of brine, and dried over anhydrous sodium sulfate. The solid is removed by filtration and the filtrate is concentrated over a rotary evaporator under reduced pressure. The residue is cooled over an ice-water bath to accelerate the crystallization. The solid is further purified by recrystallization with minimum toluene and the crystals are collected by filtration and air-dried. The product should be characterized by melting point measurement and TLC analysis. For TLC analysis, dissolve ~2 mg of the crystal in 0. 5 mL of ethyl acetate and load it onto a TLC plate. Then the plate is developed using 1 ∶ 1 EtOAc: petroleum ether as the developing solvent. After TLC development, the plate should be visualized under UV light at 254 nm.

Infrared, ultraviolet-visible and NMR spectroscopic analysis is suggested to characterize the crystal if available.

Submit the product to the instructor. Clean-up and check-out before leaving the lab.

---

① This is modified from Robert F. Nystrom, Saul W. Chaikin, Weldon G. Brown, *Journal of the American Chemical Society*, 1949, *71*(9), 3245 - 3246.

Technique Checklist

1. Addition of powder reagents in small portions.
2. Magnetic stirring.
3. Rota-evaporation.
4. Recrystallization and suction filtration.

Pre-lab Discussion

1. General mechanism of reduction by hydride transfer agents.
2. Principle of rota-evaporation.

Experiment Outline

1. Use magnetic stirring to allow the mixture to mix thoroughly.
2. Extract the product with ethyl acetate.
3. Remove volatiles with rota-evaporation.
4. Recrystallize the crude with toluene to further purify the product.
5. Analyze the purity of the product by TLC and melting point test.

Helpful Hints

1. The reaction mixture should be cooled prior to the addition of sodium borohydride in order to decrease the decomposition of sodium borohydride and also to enhance reaction selectivity.
2. The melting point of 1-(3-nitrophenyl)ethanol is 61.5 ℃.

Results

1. Record the reaction phenomena.
2. Record the mass, melting point, and TLC result for the product.
3. Calculate the isolated yield of the reaction and the R_f of the product.

5.28 Reduction of 3-Nitroacetophenone Using Sn/HCl

Objectives

1. To practice a reduction reaction using tin/acid.
2. To briefly understand the chemoselectivity of tin/acid.

Hazards

1. Chemical hazards

Granular tin, 3-nitroacetophenone, conc. hydrochloric acid, 5% methanol in dichloromethane, dichloromethane, 30% sodium hydroxide aqueous solution.

2. Physical hazards

Heating might cause skin-burning and broken glassware may lead to skin damage.

Equipment

1. 50 mL Addition funnel.
2. 250 mL Beaker.
3. Büchner funnel and suction flask.
4. Condenser.
5. 100 mL Erlenmeyer flask.
6. Filter paper and pH paper.
7. Flask tongs.
8. 50 mL Graduate cylinder.
9. Heating set-up.
10. Ice-water bath.
11. Iron ring, iron stand and clamps.
12. Magnetic stirrer with a suitable magnetic stir bar.
13. Melting point apparatus and melting tubes.
14. Pasteur pipette.
15. Rubber adapter.
16. 125 mL Separatory funnel.
17. TLC plate, TLC chamber, aluminium foil and capillary tube.
18. UV light at 254 nm.
19. Water pump.

Pre-lab work

Before entering the wet lab, read thoroughly about the experiment and understand the physicochemical principle behind. Plan your set-up of equipment by visiting useful sources and complete the pre-lab report.

Introduction

In addition to the reduction by hydride transfer agents, other reduction methods are also important. For instance, the dissolving metal reduction method is widely used. Among the many dissolving metal reduction combinations, tin/acid is widely used to reduce nitro groups into amino groups selectively, without influence on other functional groups, such as carbonyl groups in aldehydes/ketones.

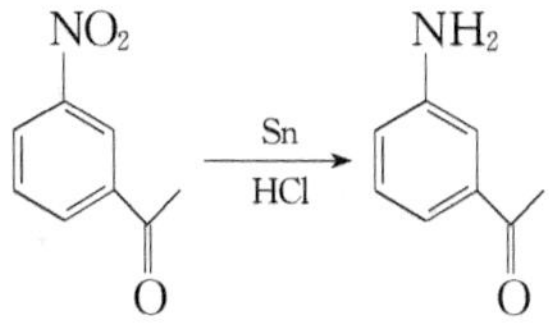

Procedure for the Sn/HCl reduction

Place 1. 65 g of 3-nitroacetophenone and 3. 3 g of granular tin in a 100 mL Erlenmeyer flask equipped with a stir bar. Attach the Erlenmeyer flask with an addition funnel containing 33 mL of conc. hydrochloric acid. The acid is dropwise added to the Erlenmeyer flask over 10 min while the reaction is stirred. After all acid is added, the addition funnel is replaced with a condenser and the reaction is heated over a 100 ℃ water bath for 30 min. Then the reaction is cooled over an ice-water bath for 10 min and is added with 30% sodium hydroxide slowly until the reaction pH reaches ~10. The resulting slurry is heated to boil for 10 min and quickly filtered and the solid is washed with 10 mL of boiled water twice. The solid is disposed into the "Solid Waste" container. The filtrate is collected, and cooled to room temperature and then in an ice-water bath to accelerate the crystallization. The crude is collected by suction filtration and the filtrate is neutralized with conc. HCl before disposal. The solid is further purified by recrystallization with water and the crystals are collected by suction filtration, washed with 4~5 mL of ice-cold water and dried over air. The product should be characterized by melting point measurement and TLC analysis. For TLC analysis, dissolve ~2 mg of the crystal or the starting material in 1 mL of dichloromethane respectively and load the solutions onto a TLC plate at different spots. Then the plate is developed using 5% methanol in dichloromethane as the developing solvent. After TLC development, the plate should be visualized under UV light at 254 nm.

Infrared, ultraviolet-visible and NMR spectroscopic analysis is suggested to characterize the crystal if available.

Submit the product to the instructor. Clean-up and check-out before leaving the lab.

Technique Checklist

1. Magnetic stirring.
2. Dropwise addition.
3. Recrystallization and suction filtration.

Pre-lab Discussion

Selectivity of reduction by tin/acid.

Experiment Outline

1. Use magnetic stirring to allow the mixture to mixture thoroughly.
2. Heat the reaction over a 100 ℃ water bath.
3. Adjusting the reaction pH to ～10.
4. Boiling the reaction and filtering off the solid while the mixture is hot.
5. Recrystallizing the crude to further purify the product.
6. Analyzing the purity of the product by TLC and melting point test.

Helpful Hints

1. Addition of conc. HCl should be slow. If foam forms, the addition should be suspended until it subsides.
2. Addition of 30% sodium hydroxide should be with caution.
3. Filtration should be rapid and the mixture should be filtered as hot as possible.
4. The flask prior to filtration is hot and flask tongs should be used to avoid burning oneself.
5. The melting point of *m*-aminoacetophenone is 92～94 ℃. ①

Results

1. Record the phenomena during the experiment
2. Record the mass, melting point, and TLC results for the product
3. Calculate the isolated yield of the reaction and R_f of the product

5.29 Synthesis of 4-Vinylbenzoic Acid via the Wittig Reaction

Objectives

1. To understand the mechanism of the Wittig reaction.
2. To prepare 4-vinylbenzoic acid via a two-step process.

Hazards

1. Chemical hazards

4-Bromomethylbenzoic acid, triphenylphosphine, acetone, diethyl ether, 37%

---

① The data was reported in Robert G. Christiansen, Raymond R. Brown, Allan S. Hay, Alex Nickon, Reuben B. Sandin, *Journal of the American Chemistry Society*, 1955, 77(44), 948-951.

formaldehyde aqueous solution, sodium hydroxide, water, ethanol, conc. HCl, dichloromethane, 50% ethyl acetate in dichloromethane.

2. Physical hazards

Broken glassware may lead to skin damage and heating might cause skin-burning.

Equipment

1. 25 mL Addition funnel.
2. 250 mL Beaker (to boil water).
3. Büchner funnel and suction flask.
4. Condenser.
5. 100 mL Erlenmeyer flask.
6. Filter paper and rubber adapter.
7. 100 mL Graduate cylinder.
8. Heating set-up.
9. Iron ring, iron stand and clamps.
10. Magnetic stirrer with a suitable magnetic stir bar.
11. Melting point apparatus and melting tubes.
12. 100 mL Round bottom flask.
13. Rubber tubing.
14. 125 mL Separatory funnel.
15. Test tubes.
16. TLC plate, TLC chamber, aluminium foil and capillary tube.
17. UV light at 254 nm.
18. Water pump.

Pre-lab work

Before entering the wet lab, read thoroughly about the experiment and understand the physicochemical principle behind. Plan your set-up of equipment by visiting useful sources and complete the pre-lab report.

Introduction

Carbon-carbon double bonds are one of the important functional moieties present in various organic compounds. It is quite often to construct carbon-carbon double bonds in a synthetic plan. Among the many olefin preparation methods, the Wittig reaction is one of the best known and most useful strategies to construct C=C from aldehyde/ketone and phosphorus ylide. This reaction was discovered and developed by Georg Wittig and is extremely chemoselective. Phosphonium ylides, phosphorus-stabilized carbanions, can be obtained by deprotonation of alkyl triphenylphosphonium salts using strong bases while the alkyl triphenylphosphonium salts can be prepared from alkyl halides and triphenylphosphine via a typical S_N2 reaction. The Wittig reaction, in principle, is a two-step reaction: An initial nucleophilic addition onto the carbonyl group to afford a betaine intermediate, which quickly undergoes rearrangement to rupture the highly strained four-membered ring and releases the newly formed olefin in addition to triphenylphosphine oxide. For example, to achieve the synthesis of *p*-vinylbenzoic acid, the ylide can be obtained by treating *p*-bromomethylbenzoic acid with nucleophilic triphenylphosphine and then react with formaldehyde in presence of bases as shown below.

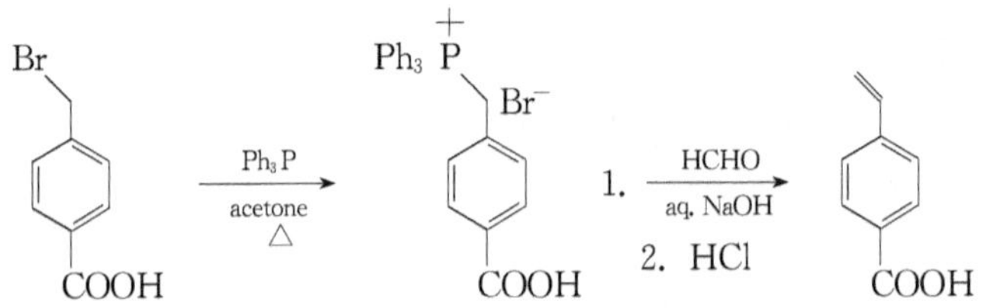

Br, COOH —(Ph_3P, acetone, △)→ $Ph_3\overset{+}{P}$, Br^-, COOH —(1. HCHO, aq. NaOH; 2. HCl)→ COOH

Procedure

1. *Preparation of the phosphonium salt*

In a 100 mL round bottom flask is dissolved 4. 3 g of 4-bromomethylbenzoic acid in 60 mL of acetone and then is added 5. 20 g of triphenylphosphine in one portion. The mixture is subsequently heated to reflux for 45 min in a hot water bath and cooled to room temperature. The solid is filtered in vacuum, washed with 2×20 mL of diethyl ether, and dried over air for 10 min. The solid is weighed and used directly for the next step.

2. *Synthesis of the olefin*

In a 100 mL Erlenmeyer flask equipped with a stir bar, 3. 7 g of the product from the previous step is mixed with 32 mL of 37% aqueous formaldehyde solution and 15 mL of water. ① The mixture is stirred vigorously over a magnetic stirrer and 2. 5 g of sodium hydroxide in 15 mL of water is added dropwise via a 25 mL addition funnel over 10 min. The resulting mixture is further stirred at room temperature for additional 45 min and the slurry is vacuum filtered to collect the solid. The pellet is washed with ~ 10 mL of water. All filtrate is combined and cooled in an ice-water bath, after which conc. hydrochloric acid is added slowly with caution to neutralize the filtrate in order to precipitate the desired product. The crude is collected by suction filtration and washed with ~ 30 mL of cold water. The solid is further

① One should adjust the amounts of reagents based on the actually phosphonium intermediate one obtains from the first step.

purified by recrystallization using 30% (v/v) aqueous ethanol solution and the crystals are collected by suction filtration, washed with 4～5 mL of ice-cold water and dried over air. The product should be characterized by melting point measurement and TLC analysis. For TLC analysis, dissolve ～2 mg of the crystal or the starting material in 1 mL of dichloromethane respectively and load the solutions onto a TLC plate at different spots. Then the plate is developed using 50% ethyl acetate in dichloromethane as the developing solvent. After TLC development, the plate should be visualized under UV light at 254 nm.

Infrared, ultraviolet-visible and NMR spectroscopic analysis is suggested to characterize the product if available.

Submit the product to the instructor. Clean-up and check-out before leaving the lab.

Technique Checklist

1. Magnetic stirring.
2. Addition of reagent dropwise using an addition funnel.
3. Recrystallization.
4. Suction filtration.
5. TLC analysis and melting point measurement.

Pre-lab Discussion

1. Mechanism for the preparation of phosphonium salts.
2. Mechanism of the Wittig reaction.

Experiment Outline

1. Prepare the phosphonium salt under refluxing.
2. Mix the newly formed phosphonium with formaldehyde in the presence of sodium hydroxide.
3. Further purify the product by recrystallization.
4. Analyze the purity of the product by TLC analysis and melting point test.

Helpful Hints

1. The alkyl triphenyl phosphonium salt can be directly used without further purification.
2. Sodium hydroxide solution is necessarily added to the Wittig reaction to dissolve the product into aqueous layer and produce the ylide for the reaction.
3. The melting point of 4-vinylbenzoic acid is 142～144 ℃.

Results

1. Record the reaction phenomena.
2. Record the mass, melting point, and TLC results for the product.
3. Calculate the isolated yield and R_f for the product.

5.30 Preparation of 1,5-Diphenylpenta-1,4-dien-3-one Using the Aldol Condensation

Objectives

1. To understand the mechanism of the aldol condensation.
2. To carry out the aldol condensation under basic conditions.

Hazards

1. Chemical hazards

Benzaldehyde, sodium hydroxide, acetone, ethanol, ethyl acetate, 10% ethyl acetate in petroleum ether.

2. Physical hazards

Heating might cause skin-burning and broken glassware may lead to skin damage.

Equipment

1. 250 mL Beaker.
2. Büchner funnel and suction flask.
3. Condenser.
4. 125 mL Erlenmeyer flask.
5. Filter paper and rubber adapter.
6. 10 mL and 25 mL Graduated cylinders.
7. Heating set-up.
8. Iron stand and clamps.
9. Magnetic stirrer with a suitable magnetic stir bar.
10. Melting point apparatus and melting capillary tubes.
11. Rubber tubing.
12. Stir rod.
13. Test tubes.
14. TLC plate, TLC chamber, aluminium foil and capillary tube.
15. UV light at 254 nm.

16. Water pump.

Pre-lab work

Before entering the wet lab, read thoroughly about the experiment and understand the physicochemical principle behind. Plan your set-up of equipment by visiting useful sources and complete your pre-lab report.

Introduction

The aldol condensation is a reaction between aldehydes or ketones under basic or acidic conditions to afford α, β-unsaturated carbonyl compounds. The reaction is useful and synthetically important when the reaction takes place between the same aldehyde molecules or between aldehyde/ketone containing α-hydrogen and aldehyde/ketone without α-hydrogen atoms. For instance, acetone can react with benzaldehyde via the aldol condensation under basic conditions, which is shown below.

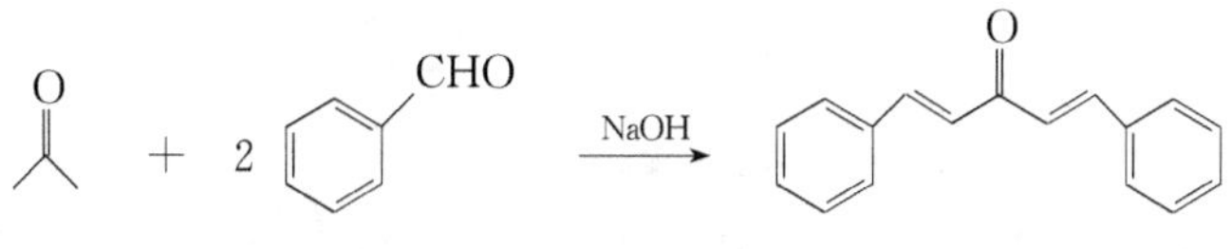

O CHO + 2 NaOH O

Procedure

To a 125 mL Erlenmeyer flask is added 2.5 g of sodium hydroxide in 13 mL of water and 13 mL of ethanol. Then 1.8 mL of acetone and 5.3 mL of benzaldehyde are added to the mixture at room temperature and the resulting mixture is stirred at room temperature for 50 min. The mixture is subject to suction filtration. The solid is washed with a small amount of cold ethanol and air-dried. The solid is further purified via recrystallization using ethyl acetate.

The product is analyzed by TLC analysis and melting point measurement. A 10% ethyl acetate solution in petroleum ether is used as the TLC developing solvent. It is also suggested that IR, UV, and NMR is used to further characterize the product following the instrumental operation instructions if available.

Submit the product to the instructor. Clean-up and check-out before leaving the lab.

Technique Checklist

1. Magnetic stirring.
2. Suction filtration.
3. Recrystallization.
4. TLC analysis.

5. Melting point measurement.

Pre-lab Discussion

Mechanism of the aldol condensation.

Experiment Outline

1. Mix sodium hydroxide, acetone, and benzaldehyde in 1 : 1 ethanol/water thoroughly.
2. Filter to collect the crude product *in vacuo*.
3. Recrystallize the crude using ethyl acetate.
4. Perform TLC and melting point analysis.

Helpful Hints

The melting point of the final product is ~113.3 ℃.

Results

1. Record the phenomena during your operation.
2. Record the amount of the crude product and recrystallized product.
3. Record the melting point of the product and the TLC result.
4. Calculate the isolated yield and R_f value of the product.

5.31 Preparation of 2-Methylbutan-2-ol via the Grignard Reagent

Objectives

1. To understand the mechanism of carbonyl nucleophilic addition of aldehydes/ketones by the Grignard reagent.
2. To practice the preparation of the Grignard reagent from corresponding alkyl halide.

Hazards

1. Chemical hazards

Granular magnesium or magnesium strip, ethyl bromide, iodine, anhydrous acetone, anhydrous diethyl ether, 20% sulfuric acid aqueous solution, anhydrous calcium chloride, anhydrous sodium carbonate

2. Physical hazards

Heating might cause skin-burning and broken glassware may lead to skin damage.

Equipment

1. 50 mL Addition funnel.
2. Distillation head and receiver adapter.
3. Drying tubes.
4. 50 mL Erlenmeyer flasks.
5. 25 mL Graduated cylinder.
6. Heating set-up.
7. Ice-water bath.
8. Iron ring, iron stand and clamps.
9. Liebig condenser.
10. Mechanical (overhead) stirrer with push-through agitator shafts.
11. Refluxing condenser.
12. Rubber tubing.
13. 250 mL Separatory funnel.
14. Thermometer with thermometer adapter.
15. 100 mL Three-neck round-bottom flask.
16. Water bath.

Pre-lab work

Before entering the wet lab, read thoroughly about the experiment and understand the physicochemical principle behind. Plan your set-up of equipment by visiting useful sources and complete the pre-lab report.

Introduction

The Grignard reagents are highly nucleophilic and can react with various electrophiles to construct carbon-carbon single bonds. The Grignard reagents can be prepared by treating the corresponding alkyl halides with elemental magnesium and typically a small quantity of iodine is added to facilitate the reaction. Because of the high basicity and nucleophilicity of organomagnesium compounds, aprotic and non-electrophilic solvents must be used for the production of organomagnesium reagents or nucleophilic reactions using the Grignard reagents.

$$\text{R—X} \xrightarrow[\text{ether}]{\text{Mg}} \text{RMgX} \qquad \text{X=Cl, Br, I}$$

Many electrophiles can react efficiently with Grignard reagents, including alkyl halides, carbonyl compounds, and epoxides, *etc.* The nucleophilic addition reaction with aldehydes or ketones is a good measure to extend the carbon chain and synthesize various secondary or tertiary alcohols. For instance, ethylmagnesium bromide can nucleophilically attack acetone, followed by an acid workup to afford the tertiary alcohol 2-methylbutan-2-ol, as shown below.

$$(CH_3)_2C{=}O + CH_3CH_2MgBr \longrightarrow (CH_3)_2C(O^-)CH_2CH_3\ \ {}^{+}MgBr \xrightarrow{H_3O^+} (CH_3)_2C(OH)CH_2CH_3$$

Procedure

1. *Preparation of ethylmagnesium bromide*

A 100 mL three-neck flask containing 3.4 g of granular magnesium and a small piece of iodine is equipped with a condenser attached with a drying tube containing anhydrous calcium chloride, a 50 mL addition funnel holding 13 mL of bromoethane in 30 mL of diethyl ether with a stopper, and a mechanical stirrer with push-through agitator shafts. The bromoethane solution is dropwise added to magnesium and the addition speed should be controlled to just allow a gentle boiling of the reaction mixture. After completion of adding the bromoethane solution, the reaction is further stirred under refluxing over a water bath for 0.5 h, until all magnesium disappears. The reaction is then cooled under an ice-water bath and the mixture is used in the following step without further treatment.

2. *Procedure for the nucleophilic addition*

The mixture of 10 mL of anhydrous acetone and 10 mL of anhydrous diethyl ether is dropwise added to the resulting solution of the above process using an addition funnel over an ice-water bath and the reaction is further stirred at room temperature for 15 min. Then the reaction vessel is cooled over an ice-water bath and 60 mL of pre-cooled 20% aqueous sulfuric acid solution is dropwise added to quench the reaction. The mixture is transferred to a separatory funnel and the organic layer is collected. The aqueous layer is further washed with 20 mL of fresh diethyl ether twice. All organic layers are combined, washed with 15 mL of 5% sodium carbonate aqueous solution, and dried over anhydrous potassium carbonate. The mixture is then filtered and the filtrate is subject to atmospheric distillation to remove the volatile first and then to collect the fraction at 95 ~ 105 ℃. The product should be further characterized by IR and NMR if available.

Technique Checklist

1. Mechanic stirring.
2. Reaction under anhydrous conditions.
3. Extraction and distillation.

Pre-lab Discussion

Mechanism of the nucleophilic addition using the Grignard reagent.

Experiment Outline

1. Use mechanic stirring to mix the reaction mixture thoroughly under anhydrous conditions.
2. Quench the reaction before further treatment.
3. Perform extraction.
4. Perform atmospheric distillation.

Helpful Hints

1. All glassware used for the preparation of the Grignard reagent or the following nucleophilic addition should be dried in the oven ahead of time.
2. Anhydrous calcium chloride in the drying tube can prevent the moisture in the air from getting in the reaction vessel.
3. The boiling point of 2-methylbutan-2-ol is 102.5 ℃.
4. 2-Methylbutan-2-ol can form an azeotrope with water in a ratio of 72.5 : 27.5 and the boiling point is ~ 87.4 ℃.
5. Addition of various reagents using the addition funnel should be slow to avoid side reactions.
6. The reaction should be slowly and carefully quenched by the dilute acid solution first.

Results

1. Record the reaction phenomena.
2. Record the mass and boiling point for the product.
3. Calculate the isolated yield of the reaction.

5.32 Preparation of Dimedone via the Robinson Annulation

Objectives

1. To understand the mechanism of the Robinson annulation and acidic decarboxylation.
2. To carry out a multi-step reaction.

Hazards

1. Chemical hazards

Sodium ethoxide, diethyl malonate, 4-methylpent-3-en-2-one, anhydrous ethanol, 60 ~ 90 ℃ petroleum ether, diethyl ether, potassium hydroxide, conc. hydrochloric acid.

2. Physical hazards

Heating might cause skin-burning and broken glassware may lead to skin damage.

Equipment

1. 50 mL Addition funnel.
2. Büchner funnel and suction flask.
3. Distillation head and receiver adapter.
4. 50 mL Erlenmeyer flasks.
5. Filter paper and pH paper.
6. 25 mL Graduated cylinder.
7. Heating set-up.
8. Ice-water bath.
9. Iron stand and clamps.
10. Liebig condenser.
11. Melting point apparatus and capillary tubes for melting point test.
12. Magnetic stirrer with a magnetic stir bar.
13. Oven
14. Refluxing condenser.
15. Rubber adapter and rubber tubing.
16. 250 mL Separatory funnel.
17. Thermometer with thermometer adapter.
18. 100 mL Three-neck round-bottom flask.

Pre-lab work

Before entering the wet lab, read thoroughly about the experiment and understand the physicochemical principle behind. Plan your set-up of equipment by visiting useful sources and complete the pre-lab report.

Introduction

There are various methods to construct cyclic compounds and one of those relies on the carbonyl chemistry is called the Robinson annulation. As a matter of fact, the

Robinson annulation is a combination of two successive reactions: the Michael addition and intramolecular carbonyl condensation reaction such as the intramolecular aldol condensation or Dieckmann condensation. Via such reactions, six-membered rings are usually formed. For instance, in this experiment, diethyl malonate, as an active methylene compound, can be easily converted into a good nucleophile by the treatment of base, which nucleophilically attacks the α,β-unsaturated ketone via a 1, 4-addition mechanism to achieve a 1,5-dicarbonyl compound. While the large excess of base is able to remove the α-hydrogen of the ketone residue in the 1,5-dicarbonyl compound, a Dieckmann condensation takes place to close up the cycle and form a six-membered ring. Following a saponification and an acid treated decarboxylation, the final product dimedone is obtained.

O
+
EtO_2C CO_2Et
NaOEt
EtOH
△
CO_2Et
O O
⊖
$Na^{\oplus}$
1. NaOH, H_2O
2. H_3O^+, △
O O
Dimedone

Procedure

Under magnetic stirring, 7.6 mL of diethyl malonate is dropwise added to 16 mL of 25% sodium ethoxide in anhydrous ethanol over 5 min in a 100 mL oven-dried three-neck flask①. Then 5 mL of anhydrous ethanol is added to rinse the addition funnel into the reaction. After this, 5.7 mL of 4-methylpent-3-en-2-one is added dropwise to the mixture followed by another 5 mL of anhydrous ethanol. The reaction is stirred and heated to reflux gently for 45 min. Subsequently, a solution containing 6.3 g of potassium hydroxide in 25 mL of water is slowly added to the reaction through the addition funnel. Upon the completion of the addition, the mixture is further refluxed for additional 45 min and then cooled a bit to allow a distillation set-up to be built for the reaction. The reaction mixture is concentrated by removing around 35 mL of the azeotrope of ethanol and water via atmospheric distillation, after which the residue is cooled over an ice-water bath. The cooled residue is then extracted with 25 mL of diethyl ether and the aqueous layer is collected. The diethyl ether layer is further washed with 2×10 mL of water and the aqueous layers are combined with the initial aqueous fraction. The aqueous solution is acidified with conc. hydrochloric acid to pH ~1 and heated to reflux for 15 min. Afterwards, the reaction mixture is cooled over an ice-water bath until no more crystals form. The

① All glassware and the magnetic stir-bar should be dried in the oven before use for the reaction.

crystals are collected by suction filtration, thoroughly washed with 25 mL of water and then 25 mL of petroleum ether, and dried. Recrystallize the crude with water to give the final product. The melting point of the product is measured and the product should be further characterized by IR and NMR if available.

Technique Checklist

1. Magnetic stirring.
2. Reaction under anhydrous and refluxing conditions.
3. Dropwise addition of reagents to a reaction.
4. Distillation and extraction.
5. Suction filtration and recrystallization.

Pre-lab Discussion

1. Mechanism of the Robinson annulation.
2. Mechanism of the saponification and decarboxylation.

Experiment Outline

1. Dropwise add a reagent to the reaction.
2. Stir and heat the reaction to reflux.
3. Perform atmospheric distillation to concentrate the reaction.
4. Extrac the aqueous residue with diethyl ether.
5. Acidify the aqueous layer and heat to reflux.
6. Crystallize and filter the product.
7. Recrystalize the solid with water.

Helpful Hints

1. The glassware used for the first step of the reactions should be dried in the oven ahead of time.
2. The melting point of dimedone is 148～149 ℃.
3. Completely mixing the reaction mixture is important for the reaction.

Results

1. Record the reaction phenomena.
2. Record the mass of the crude and the final product.
3. Record the melting point for the product.
4. Calculate the isolated yield of the reaction.

5.33 Preparation of Phenacetin from *p*-Aminophenol via a Two-Step Synthesis

Objectives

1. To understand the synthetic route for phenacetin.
2. To appreciate the power of organic synthesis in drug development.
3. To comprehensively apply basic synthetic techniques.

Hazards

1. Chemical hazards

Acetic anhydride, *p*-aminophenol, sodium ethoxide, iodoethane, 20% petroleum ether in diethyl ether.

2. Physical hazards

Heating might cause skin-burning and broken glassware may lead to skin damage.

Equipment

1. Büchner funnel and suction flask.
2. Condenser.
3. 50 mL Erlenmeyer flask.
4. Filter paper and rubber adapter.
5. 10 mL Graduated cylinders.
6. Heating set-up.
7. Ice-water bath.
8. Iron stand and clamps.
9. Magnetic stirrer with a suitable magnetic stir bar.
10. Melting point apparatus and melting tube.
11. Oven.
12. Pasteur pipette.
13. 50 mL Round bottom flask.
14. Stir rod.
15. TLC plate, TLC chamber, aluminium foil and capillary tube.
16. UV light at 254 nm.
17. Water pump.

Pre-lab work

Before entering the wet lab, read thoroughly about the experiment and understand the physicochemical principle behind. Plan your set-up of equipment by visiting useful sources and complete your pre-lab report.

Introduction

Along the development of organic synthesis, there have been tremendous amounts of natural compounds prepared in research laboratories and industrial plants. Meanwhile, many new techniques and organic reactions have also been developed. For instance, phenacetin was an analgesic and fever reducing drug used for human beings and animals since its introduction in 1887. However, it was found to be associated with kidney disease and later carcinogenic effects, so that it was withdrawn from the market since 1978. The metabolic pathways for phenacetin include *de*-ethylation to afford paracetamol, *N*-*de*acetylation to give *p*-phentidine, or hydroxylation to introduce hydroxy groups onto the benzene ring.

HO–C₆H₄–NH_2 $\xrightarrow{(CH_3CO)_2O}$ Paracetamol $\xrightarrow[NaOEt]{EtI}$ Phenacetin

Typically, phenacetin can be prepared via two steps from *p*-aminophenol. The first step usually involves a classic *N*-acetylation. As the other product acetic acid may react with the amino group and decrease its nucleophilicity, a base, such as pyridine or 4-pyrrolidinopyridine, is always added to facilitate the reaction. The second step is a Williamson ether synthesis, which is efficient for phenyl ether preparation. To drive the reaction forward, the phenol is first deprotonated by a base, such as sodium ethoxide to generate a good nucleophile, phenoxide, which can react rapidly with iodoethane through an S_N2 mechanism with high efficiency. It is worth noting that the reaction sequence cannot be switched to alkylation and then acetylation, which may lead to severe side products and decrease the overall yield of the target molecule.

Procedure

1. *Acetylation*

To a 50 mL Erlenmeyer flask with a magnetic stir bar are added 2.0 g of *p*-aminophenol, 2.2 mL of acetic anhydride, and 10 mL of water. The mixture is stirred and heated to nearly boil for 30 min. Then the reaction is cooled over an ice-water bath and filtered in vacuum. The solid is further washed with a small amount of cold water and recrystallized with water. The recrystallization flask is further cooled over an ice-water bath to further precipitate the product after it is cooled to room temperature. The crystal is collected via suction filtration, washed with a small amount of cold water, and dried in an oven. ①

2. *Alkylation*

Mix 1.7 g of the product from the first step and 0.82 g of sodium ethoxide in 10 mL of ethanol in a 50 mL round bottle flask and 1.4 mL of iodoethane is added drop by drop over 1 min. The mixture is then heated under refluxing for 1 h. The reaction is cooled over an ice-water bath and 10 mL of water is carefully added to quench the reaction. The crude is collected via suction filtration. The product is further purified by recrystallization with water and dried in air. Products from the two steps should be analyzed by TLC and melting point measurement. A 20% Petroleum ether solution in diethyl ether is used as the TLC developing solvent. It is suggested that IR, UV, and NMR is used to further characterize the product if available.

Submit the product to the instructor. Clean-up and check-out before leaving the lab.

Technique Checklist

1. A reaction under refluxing.
2. Suction filtration.
3. Recrystallization.
4. TLC analysis of the product.
5. Measurement of the melting point of the product.

Pre-lab Discussion

1. Mechanism of both reactions.
2. Logic of the reaction sequence.

Experiment Outline

1. Mix and heat *p*-aminophenol with acetic anhydride in water to nearly boil.
2. Suction filtration to collect the crude product of the first reaction.
3. Recrystallization using water.
4. Heat paracetamol, iodoethane, and sodium ethoxide in ethanol to reflux.
5. Suction filtration to collect the crude product of the second reaction.

---

① If the oven is not available in the undergraduate laboratory, the sample should be spread on a watch glass and carefully baked on a hot plate with a gentle heating.

6. TLC and melting point analysis.

Helpful Hints

1. The melting points of paracetamol and phenacetin are 169 ℃ and 134 ℃ respectively.

2. Dry glassware should be used for both reactions.

Results

1. Record the phenomena during the operation.
2. Record the amount of the crude products and recrystallized products.
3. Record the melting point of the products and TLC results.
4. Calculate the overall yield of the reactions and R_f of the products.

5.34 Extraction of Alkaloids from Chinese Herbs—*Coptis Chinensis*

Objectives

1. To use the Soxhlet extractor to isolate alkaloids from Chinese herbs.
2. To understand the principle of the Soxhlet extraction.

Hazards

1. Chemical hazards

Coptis chinensis, ethanol, 1% acetic acid aqueous solution, 6 mol · L^{-1} hydrochloric acid, 10% methanol in chloroform.

2. Physical hazards

Heating might cause skin-burning and broken glassware may lead to skin damage.

Equipment

1. Alumina TLC plate, TLC chamber, aluminium foil and capillary tube.
2. 250 mL Beaker.
3. Büchner funnel and suction flask.
4. Condenser.
5. Distillation head, Liebig condenser, and receiver adapter.
6. 150 mL Erlenmeyer flask.
7. Filter paper and rubber adapter.
8. 100 mL Graduated cylinder.

9. Heating set-up.
10. Iron stand and clamps.
11. Melting point apparatus and melting tubes.
12. Pasture pipette.
13. 250 mL Round bottom flask.
14. Rubber tubing.
15. 100 mL Soxhlet extractor with a filter paper.
16. Thermometer and adapter.
17. UV light at 254 nm.
18. Water pump.

Pre-lab work

Before entering the wet lab, read thoroughly about the experiment and understand the physicochemical principle behind. Plan your set-up of equipment by visiting useful sources and complete your pre-lab report.

Introduction

Nature is the key source for a variety of biologic compounds. For instance, the fascinating effect of Chinese herbs relies on the effective components present in the herbs. However, there are a lot of compounds present in the herbs, some of which may cause unwanted side effects. Therefore, it is critical to obtain the effective component and may as well identify its structure and thereafter get a better understanding of its pharmacodynamics and pharmacokinetics. It would also be possible for medicinal scientists to modify the structures to pinpoint the key structures and functional groups that play significant roles in its effect.

There are many ways to isolate natural compounds. Dissolution is one key method, among which the liquid-solid extraction is tremendously used to extract the organic compounds from the source into organic solutions. The Soxhlet extraction provides a feasible way to use a small amount of solvent to continuously extract compounds from the solid into the liquid solution in combination of liquid-solid extraction and distillation-condensation in the system.

Berberine

Berberine is an alkaloid and the key effective component responsible for the high antimicrobial properties in coptis chinensis. Berberine is also found to be effective in pain killing, and cholesterol and sugar reducing, *etc.* Berberine is generally a quaternary ammonium salt and its structure is shown beside. It can be extracted using the liquid-solid extraction method as it demonstrates good thermostability and some solubility in organic solvents.

Procedure

Coptis chinensis (5 g) is cut into small pieces, grinded, and placed in the extraction thimble of a 100 mL Soxhlet extractor. The extractor is attached with a condenser and a 250 mL round bottom flask containing 110 mL of ethanol. The round bottom flask is heated to reflux until the liquid in the extraction thimble is faint. Then the Soxhlet extractor is removed and the solution is concentrated by atmospheric distillation. The residue is added with ~40 mL of 1% acetic acid solution and heated to boil for 5 min. The mixture is filtered and the filtrate is cooled to room temperature and acidified with 6 mol • L^{-1} hydrochloric acid until its pH reaches 1~2. The mixture is further cooled in an ice-water bath for 5 min. The solid is collected by suction filtration and dried over 50~60 ℃ in the oven. The solid is then analyzed by melting point measurement and TLC analysis. Approximately ~2 mg of the solid is dissolved in 1 mL ethanol and loaded on an alumina TLC plate. The TLC plate is then developed in a TLC chamber using 10% methanol in chloroform as the developing solvent. After the TLC development, the plate should be visualized under UV light at 254 nm.

Submit the product to the instructor. Clean-up and check-out before leaving the lab.

Technique Checklist

1. Liquid-solid extraction using the Soxhlet extractor.
2. Atmospheric distillation to concentrate a solution.
3. Acidification of the residual solution.
4. Suction filtration.
5. TLC analysis of the final product.

Pre-lab Discussion

Principle of the Soxhlet extraction.

Experiment Outline

1. Cut and grind the herb into small particles.
2. Perform soxhlet extraction using ethanol.
3. Concentrate the solution using atmospheric distillation.
4. Acidify and precipite the residue.
5. Undertake Suction filtration.

6. Perform TLC analysis and melting point measurement.

Helpful Hints

1. The Soxhlet extraction takes around 2.5 h.

2. If a regular silica TLC plate is used, the developing solvent should be 7 : 2 : 1 butan-1-ol : water : ethanol.

Results

1. Record the phenomena during your operation.
2. Record the amount of the crude product.
3. Record the melting point of the product.
4. Record the result of TLC.
5. Calculate the isolated yield and R_f of the product.

5.35 Extraction of Caffeine from Teas

Objectives

1. To use dissolution and extraction to transfer caffeine from tea to the organic solution.

2. To purify caffeine via recrystallization.

Hazards

1. Chemical hazards

Tea, water, sodium carbonate, dichloromethane, 95% ethanol, Celite®.

2. Physical hazards

Heating might cause skin-burning and broken glassware may lead to skin damage.

Equipment

1. 250 mL and 50 mL Beaker.
2. Büchner funnel and suction flask.
3. 250 mL Erlenmeyer flask.
4. Filter paper and rubber adapter.
5. 100 mL Graduated cylinder.
6. Heating set-up.
7. Iron ring, iron stand and clamps.
8. Melting point apparatus and melting tubes.

9. Petri dish.
10. Rota-evaporator.
11. Rubber tubing.
12. 250 mL Separatory flask.
13. Water pump.

Pre-lab work

Before entering the wet lab, read thoroughly about the experiment and understand the physicochemical principle behind. Plan your set-up of equipment by visiting useful sources and complete your pre-lab report.

Introduction

Caffeine is a type of methylxanthine, which is regarded as one of the alkaloids. It is present in tea or coffee and is the stimulant for the improvement in fatigue and concentration, *etc.* In addition to caffeine, tea or coffee contains other important biocompounds such as proteins, saccharides, pigments, vitamins, and polyphenol, *etc.*, some of which are water insoluble. These type of compounds can be achieved by isolation from tea or coffee or synthesis in the organic laboratories.

Procedure

In a 250 mL Erlenmeyer flask, 15 g of tea is mixed with 10 g of sodium carbonate and 175 mL of water. The mixture is then heated to boil for ~ 10 min, cooled under an ice-water bath, and filtered through a Celite® via suction filtration. The filtrate is transferred to a 250 mL separatory funnel and extracted with 35 mL of dichloromethane four times. The organic layers are combined and dried over anhydrous sodium sulfate for no less then 10 min. The solid is removed by filtration and the filtrate is concentrated by rotary evaporation *in vacuo*. The solid is weighed and recrystallized from 95% ethanol. The product is collected via suction filtration and air-dried. The solid is then transferred to a beaker. A Petri dish is placed on the beaker. The beaker is gently heated until no further solid disappears. Collect the solid on the Petri dish and weigh the mass. The melting point of the product is measured and it is suggested the product is further characterized using MS spectrometry, IR and NMR spectroscopy if available.

Submit the product to the instructor. Clean-up and check-out before leaving the lab.

Technique Checklist

1. Dissolution.

2. Extraction using dichloromethane.
3. Rota-evaporation under vacuum.
4. Recrystallization using 95% ethanol.
5. Suction filtration.
6. Melting point measurement.
7. Sublimation.

Pre-lab Discussion

Purification of caffeine from other compounds present.

Experiment Outline

1. Cook tea and sodium carbonate in water.
2. Extract the aqueous solution with dichloromethane.
3. Concentrate the solution by rota-evaporation under vacuum.
4. Recrystallize the solid.
5. Perform the sublimation.
6. Measure the melting point.

Helpful Hints

Typically, the solubility of caffeine is 5 mL of ethanol per gram of caffeine.

Results

1. Record the phenomena during the operation.
2. Record the amount of the crude product and purified product.
3. Record melting point of the product.
4. Calculate the isolated yield.

6 Appendices

6.1 List of Equipment and Glassware for Each Experiment

5.1 Purification of Liquids by Distillation

- Boiling chips.
- Distillation kits (distillation head, Liebig condenser, receiver adapter, thermometer with adapter).
- 50 mL Erlenmeyer flasks.
- 10 mL or 50 mL Graduated cylinder.
- Glass wool and aluminium foil (optional).
- Heating set-up.
- Iron stand and clamps.
- 25 mL & 50 mL Round bottomed flasks.
- Rubber tubing.

5.2 Purification of Organic Compounds by Extraction

- 50 mL Burette, Burette clamp, and support stand.
- 50 mL Erlenmeyer flasks.
- 50 mL Graduated cylinder.
- Iron ring, iron stand and clamps.
- 50 mL or 125 mL Separatory funnel (sep funnel).
- 5 mL and 10 mL Transfer pipettes.

5.3 Purification of Solids by Recrystallization

- Boiling chips.
- Büchner funnel and filter flask with a rubber adapter.
- Condenser.
- 250 mL Erlenmeyer flask with ground glass joint.
- Filter paper.
- 100 mL Graduated cylinder.

- Heating set-up.
- Iron stand and clamps.
- Rubber tubing.
- Stir rod or spatula.
- TLC plate and capillary tube.
- TLC chamber and aluminium foil.
- UV light (254 nm).
- Water aspirator or water pump.
- Weighing paper.

5.4 Purification of *Cyclo*hexane

- Distillation kit (distillation head, Liebig condenser, receiver adapter, thermometer with adapter).
- 50 mL Erlenmeyer flasks.
- 25 mL Graduated cylinder.
- Heating set-up.
- Iron ring, iron stand and clamps.
- Liquid funnel and filter paper.
- 50 mL Round bottom flask.
- Rubber tubing.
- 50 mL or 125 mL Separatory funnel.
- 5 mL Test tubes and Pasteur pippetes.

5.5 Preparation of Bromoethane

- Boiling chips.
- Distillation kit (distillation head, thermometer and adapter, Liebig condenser, and receiver adapter).
- 100 mL or 50 mL Erlenmeyer flasks.
- 10 mL or 25 mL Graduated cylinder.
- Heating set-up.
- Ice-water bath.
- Iron ring, iron stand and clamps.
- Pasteur pipettes.
- 100 mL Round bottom flask.
- Rubber tubing.
- 50 mL or 125 mL Separatory funnel (sep-funnel).
- 10 mL Transfer pipettes.

5.6 Synthesis of Ethyl Acetate

- Adapters.
- Boiling chips.
- Condenser.
- Distillation kit (distillation head, thermometer and adapter, Liebig condenser, and receiver adapter).
- 10 mL Graduated cylinder.
- Heating set-up.
- Iron ring, iron stand and clamps.
- Liquid funnel and filter paper.
- Litmus paper.
- 50 mL Round bottom flasks.
- Rubber tubing.
- 125 mL or 250 mL Separatory funnel.

5.7 Synthesis of Benzoin

- Boiling chips.
- Büchner funnel and filtration flask.
- Condenser.
- Filter paper and rubber adapter.
- 10 mL, 25 mL or 50 mL Graduated cylinder.
- Heating set-up.
- Ice-water bath.
- Iron stand and clamps.
- Melting point apparatus and capillary tubes for melting point test.
- Pasteur pipette.
- 50 mL Round bottom flask.
- Rubber tubing.
- 10 mL Test tubes.
- TLC plates and capillary tubes.
- TLC chamber and aluminium foil.
- UV light at 254 nm.
- Water pump.

5.8 Synthesis of Cinnamic Acid by the Perkin Reaction

- 250 mL Beaker.
- Büchner funnel and filtration flask.

- Condensers.
- Congo red paper.
- Distillation head and receiver adapter.
- 100 mL Erlenmeyer flask.
- Filter paper and rubber adapter.
- 10 mL, 25 mL or 50 mL Graduated cylinder.
- Heating set-up.
- Iron ring, iron stand and clamps.
- Liquid funnel and filter paper.
- Melting point apparatus and capillary tubes for melting point test.
- Rubber tubing.
- 125 mL Separatory funnel with a grounded joint.
- Stirring rod.
- 5 mL Test tube.
- Thermometer and adapter.
- TLC plates and capillary tubes.
- TLC chamber and aluminium foil.
- 50 mL Two-neck round bottom flask.
- UV light at 254 nm.
- Water pump.

5.9 Synthesis of Cinnamic Acid by the Knoevenagel Condensation

- Büchner funnel and suction flask.
- Condensers.
- Filter paper and rubber adapter.
- 10 mL, 25 mL or 50 mL Graduated cylinder.
- Heating set-up.
- Iron ring, iron stand and clamps.
- Melting point apparatus and capillary tubes for melting point test.
- 50 mL Round bottom flask.
- Rubber tubing.
- 50 mL Separatory funnel.
- TLC plates and capillary tubes.
- TLC chamber and aluminium foil.
- UV light at 254 nm.
- Water pump.

5.10 Synthesis of Methyl Orange

- 50 mL & 100 mL Beakers.
- Büchner funnel and filter paper.
- Filtration flask and rubber adapter.
- 25 mL & 10 mL Graduated cylinders.
- Heating set-up.
- Ice-water bath.
- Iron stand and clamps.
- Litmus test paper or pH test paper.
- Pasteur pipettes.
- Stir rod.
- 10 mL and 5 mL Test tubes.
- Water pump.

5.11 Synthesis of Ethyl Benzoate

- 100 mL Beaker.
- Boiling chips.
- 25 mL Graduated cylinder.
- Dean-Stark apparatus.
- Distillation kit (distillation head, thermometer with adapter, Liebig condenser, receiver adapter).
- 50 mL Erlenmeyer flasks.
- Condenser.
- Heating set-up.
- Iron ring, iron stand and clamps.
- pH test papers.
- 100 mL & 50 mL Round bottom flasks.
- Rubber tubing.
- 50 mL or 125 mL Separatory funnel.
- Stir rod.
- TLC chamber and aluminium foil.
- TLC plate and capillary tubes.
- UV light at 254 nm.

5.12 The Diels-Alder Reaction

- Büchner funnel and filtration flask.
- Condenser.

- 50 mL Erlenmeyer flask.
- Filter paper and rubber adapter.
- 50 mL Graduated cylinder.
- Heating set-up.
- Ice-water bath.
- Iron stand and clamps.
- Melting point apparatus and capillary tubes for melting point test.
- 50 mL Round bottom flask.
- Rubber tubing.
- TLC chamber and aluminium foil.
- TLC plate and capillary tubes.
- UV light (254 nm).
- Water pump.

5.13 Synthesis of Acetanilide

- 500 mL Beaker.
- Boiling chips.
- Büchner funnel and filtration flask.
- Condenser.
- Distillation head.
- Filter paper and rubber adapter.
- 25 mL Graduated cylinder.
- Heating set-up.
- Iron stand and clamps.
- Melting point apparatus and capillary tubes for melting point test.
- Receiver adapter.
- 50 mL Round bottom flask.
- Rubber tubing.
- 15 mL Test tubes.
- TLC chamber and aluminium foil.
- TLC plate and capillary tubes.
- Vigreux distillation column.
- UV light at 254 nm.
- Water pump.

5.14 Bromination of Acetanilide

- Büchner funnel and filtration flask.
- Condenser.

- 2 mL Disposable syringe with a detachable needle.
- 50 mL Erlenmeyer flask.
- Filter paper and rubber adapter.
- 10 mL Graduated cylinder.
- Heating set-up.
- Iron stand and clamps.
- Magnetic stirrer with a suitable magnetic stir bar.
- Melting point apparatus and capillary tubes for melting point test.
- Rubber tubing.
- TLC chamber and aluminium foil.
- TLC plate and capillary tubes.
- UV light at 254 nm.
- Water pump.

5.15 Deprotection of the Acetyl Protecting Group from *p*-Bromoacetanilide

- Büchner funnel and filtration flask.
- Condenser.
- 100 mL Erlenmeyer flask.
- Filter paper and rubber adapter.
- 10 mL Graduated cylinder.
- Heating set-up.
- Iron stand and clamps.
- Melting point apparatus and capillary tubes for melting point test.
- 50 mL Round bottom flask.
- Rubber tubing.
- Stir rod.
- TLC chamber and aluminium foil.
- TLC plate and capillary tubes.
- UV light at 254 nm.
- Water pump.

5.16 Synthesis of Aspirin

- 100 mL Beaker.
- Büchner funnel and suction flask.
- Condenser.
- 10 mL and 50 mL Graduate cylinders.
- Filter paper and rubber adapter.
- Heating set-up.

- Ice-water bath.
- Iron stand and clamps.
- Melting point apparatus and capillary tubes for melting point test.
- pH test paper.
- 50 mL Round bottom flask.
- Rubber tubing.
- Spatula.
- Test tubes.
- TLC chamber and aluminium foil.
- TLC plate and capillary tube.
- UV light at 254 nm.
- Water pump.

5.17 Synthesis of Ferrocene via the Phase Transfer Catalysis

- 100 mL Beaker.
- Büchner funnel and suction flask.
- 50 mL Erlenmeyer flask.
- Filter paper and rubber adapter.
- Glass rod.
- 50 mL Graduated cylinder.
- Iron stand and clamps.
- Magnetic stirrer and a suitable magnetic stir bar.
- Melting point apparatus and capillary tubes for melting point test.
- Test tubes.
- TLC chamber and aluminium foil.
- TLC plate and capillary tube.
- UV light at 254 nm.
- Water pump.

5.18 Synthesis of 2-Nitro-1,3-benzenediol

- 100 mL Beaker.
- Büchner funnel and suction flask.
- Condenser.
- Distillation head and receiver adapter.
- 25 mL and 100 mL Erlenmeyer flasks.
- Filter paper and rubber adapter.
- 10 mL and 25 mL Graduated cylinders.
- Heating set-up.

- Ice-water bath.
- Iron stand and clamps.
- Magnetic stirrer and magnetic stir bar.
- Melting temperature apparatus and capillary tube.
- Mortar and pestle.
- Pasteur pipettes.
- Rubber septum and glass tubes.
- Rubber tubing.
- Separatory funnel with a ground joint.
- Stirring rod.
- Test tubes.
- Thermometer with adapter.
- TLC plate and capillary tube.
- TLC chamber and aluminium foil.
- 100 mL Two-neck round-bottom flask.
- UV light at 254 nm.
- Water bath.
- Water pump.

5.19 Preparation of the Urea-Formaldehyde Resin

- Condenser.
- 10 mL and 50 mL Graduated cylinders.
- Heating set-up.
- Iron stand and clamps.
- Mechanical (overhead) stirrer with push-through agitator shafts.
- Pasteur pipette.
- Rubber tubing.
- Stirring rod.
- Thermometer and adapter.
- 250 mL Three-neck round-bottom flask.
- Two wood blocks with clean and smooth surface.

5.20 Preparation of *Cyclo*hexene

- Distillation head.
- 50 mL Erlenmeyer flask.
- 10 mL Graduate cylinder.
- Heating set-up.
- Ice-water bath.

- Iron ring, iron stand and clamps.
- Liebig Condenser.
- Pasteur pipette.
- Receiver adapter.
- 50 mL Round bottom flask.
- Rubber tubing.
- 125 mL Separatory funnel.
- Thermometer with adapter.
- Vigreux distillation column.

5.21 Synthesis of 2-Chloro-2-methylpropane

- 500 mL Beaker.
- Distillation head and receiver adapter.
- 50 mL Erlenmeyer flask.
- 10 mL Graduated cylinder.
- Heating set-up.
- Iron ring, iron stand and clamps.
- Liebig condenser.
- Magnetic stirrer with a suitable magnetic stir bar.
- 50 mL Round bottom flask.
- Reflux condenser.
- Rubber tubing.
- 125 mL Separatory funnel.
- Thermometer with thermometer adapter.

5.22 Synthesis of Dibutyl Ether

- Boiling chips.
- Dean-Stark apparatus.
- Distillation head and receiver adapter.
- 100 mL or 50 mL Erlenmeyer flasks.
- 10 mL or 25 mL Graduated cylinder.
- Heating set-up.
- Ice-water bath.
- Iron ring, iron stand and clamps.
- Liebig condenser.
- Oil bath.
- 50 mL Round bottom flask.
- Rubber tubing.

- 125 mL Separatory funnel.
- Thermometer and adapter.
- 50 mL Three-neck flask.

5.23 Synthesis of Butyl Phenyl Ether

- Air condenser.
- Dean-Stark apparatus.
- Distillation head and receiver adapter.
- 100 mL or 50 mL Erlenmeyer flasks.
- 25 mL Graduated cylinder.
- Heating set-up.
- Iron ring, iron stand and clamps.
- Liebig condenser.
- Mechanical (overhead) stirrer with push-through agitator shafts.
- 50 mL Round bottom flask.
- Rubber tubing.
- 125 mL Separatory funnel.
- Test tubes.
- Thermometer and adapter.
- 100 mL Three-neck round-bottom flask.
- TLC chamber and aluminium foil.
- TLC plate and capillary tubes.
- UV light at 254 nm.

5.24 Synthesis of Hexane-1,6-dionic Acid

- Büchner funnel and suction flask.
- 10 mL and 50 mL Graduated cylinder.
- 250 mL Beaker.
- 250 mL Erlenmeyer flask with a magnetic stir bar.
- Filter paper and pH paper.
- Heating set-up.
- Ice-water bath.
- Iron stand and clamps.
- Magnetic stirrer.
- Melting point apparatus and melting tube.
- Reflux condenser.
- Rubber adapter.
- Stirring rod.

- Water bath.
- Water pump.

5.25 Preparation of Furfuryl Alcohol and 2-Furoic Acid by the Cannizzaro Reaction

- 50 mL Beaker.
- Büchner funnel and suction flask.
- Distillation head, Liebig condenser, and receiver adapter.
- 50 mL Erlenmeyer flasks.
- Filter paper and pH paper.
- Glass rod.
- 25 mL Graduate cylinder.
- Heating set-up.
- Ice-water bath.
- Iron ring, iron stand and clamps.
- Magnetic stirrer and a suitable magnetic stir bar.
- Melting point apparatus and melting tubes.
- Pasteur pipette.
- 100 mL Round bottom flask.
- Rubber adapter.
- 125 mL Separatory funnel.
- Test tubes.
- Thermometer with an adapter.
- TLC plate, TLC chamber, aluminium foil and capillary tube.
- UV light at 254 nm.
- Water pump.

5.26 A Solvent Free Cannizzaro Reaction

- 50 mL Beaker.
- Büchner funnel and suction flask.
- Filter paper and pH paper.
- Glass rod.
- 10 mL Graduate cylinder.
- Iron stand and clamps.
- Magnetic stirrer and a suitable magnetic stir bar.
- Melting point apparatus and melting tubes.
- Pasteur pipette.
- Pestle and mortar.

- Rubber adapter.
- Test tubes.
- TLC plate, TLC chamber, aluminium foil and capillary tube.
- UV light at 254 nm.
- Water pump.

5.27 Reduction of 3-Nitroacetophenone using Sodium Borohydride

- Büchner funnel and suction flask.
- Condenser.
- 25 mL Graduate cylinder.
- Filter paper and rubber adapter.
- Heating set-up.
- Ice-water bath.
- Iron ring, iron stand and clamps.
- Magnetic stirrer with a suitable magnetic stir bar.
- Melting point apparatus and melting tubes.
- Rota-evaporator with a chilling circulation system.
- 50 mL Round bottom flask.
- Separatory funnel.
- TLC plate, TLC chamber, aluminium foil and capillary tube.
- UV light at 254 nm.
- Water pump.

5.28 Reduction of 3-Nitroacetophenone Using Sn/HCl

- 50 mL Addition funnel.
- 250 mL Beaker.
- Büchner funnel and suction flask.
- Condenser.
- 100 mL Erlenmeyer flask.
- Filter paper and pH paper.
- Flask tongs.
- 50 mL Graduate cylinder.
- Heating set-up.
- Ice-water bath.
- Iron ring, iron stand and clamps.
- Magnetic stirrer with a suitable magnetic stir bar.
- Melting point apparatus and melting tubes.
- Pasteur pipette.

- Rubber adapter.
- 125 mL Separatory funnel.
- TLC plate, TLC chamber, aluminium foil and capillary tube.
- UV light at 254 nm.
- Water pump.

5.29 Synthesis of 4-Vinylbenzoic Acid via the Wittig Reaction

- 25 mL Addition funnel.
- 250 mL Beaker (to boil water).
- Büchner funnel and suction flask.
- Condenser.
- 100 mL Erlenmeyer flask.
- Filter paper and rubber adapter.
- 100 mL Graduate cylinder.
- Heating set-up.
- Iron ring, iron stand and clamps.
- Magnetic stirrer with a suitable magnetic stir bar.
- Melting point apparatus and melting tubes.
- 100 mL Round bottom flask.
- Rubber tubing.
- 125 mL Separatory funnel.
- Test tubes.
- TLC plate, TLC chamber, aluminium foil and capillary tube.
- UV light at 254 nm.
- Water pump.

5.30 Preparation of 1,5-Diphenylpenta-1,4-dien-3-one Using the Aldol Condensation

- 250 mL Beaker.
- Büchner funnel and suction flask.
- Condenser.
- 125 mL Erlenmeyer flask.
- Filter paper and rubber adapter.
- 10 mL and 25 mL Graduated cylinders.
- Heating set-up.
- Iron stand and clamps.
- Magnetic stirrer with a suitable magnetic stir bar.
- Melting point apparatus and melting capillary tubes.

- Rubber tubing.
- Stir rod.
- Test tubes.
- TLC plate, TLC chamber, aluminium foil and capillary tube.
- UV light at 254 nm.
- Water pump.

5.31 Preparation of 2-Methylbutan-2-ol via the Grignard Reagent

- 50 mL Addition funnel.
- Distillation head and receier adapter.
- Drying tubes.
- 50 mL Erlenmeyer flasks.
- 25 mL Graduated cylinder.
- Heating set-up.
- Ice-water bath.
- Iron ring, iron stand and clamps.
- Liebig condenser.
- Mechanical (overhead) stirrer with push-through agitator shafts.
- Refluxing condenser.
- Rubber tubing.
- 250 mL Separatory funnel.
- Thermometer with thermometer adapter.
- 100 mL Three-neck round-bottom flask.
- Water bath.

5.32 Synthesis of Dimedone via the Robinson Annulation

- 50 mL Addition funnel.
- Büchner funnel and suction flask.
- Distillation head and receiver adapter.
- 50 mL Erlenmeyer flasks.
- Filter paper and pH paper.
- 25 mL Graduated cylinder.
- Heating set-up.
- Ice-water bath.
- Iron stand and clamps.
- Liebig condenser.
- Melting point apparatus and capillary tubes for melting point test.
- Magnetic oven stirrer with a magnetic stir bar.

- Refluxing condenser.
- Rubber adapter and rubber tubing.
- 250 mL Separatory funnel.
- Thermometer with thermometer adapter.
- 100 mL Three-neck round-bottom flask.

5.33 Preparation of Phenacetin from *p*-Aminophenol via a Two-Step Synthesis

- Büchner funnel and suction flask.
- Condenser.
- 50 mL Erlenmeyer flask.
- Filter paper and rubber adapter.
- 10 mL Graduated cylinders.
- Heating set-up.
- Ice-water bath.
- Iron stand and clamps.
- Magnetic stirrer with a suitable magnetic stir bar.
- Melting point apparatus and melting tube.
- Oven.
- Pasteur pipette.
- 50 mL Round bottom flask.
- Stir rod.
- TLC plate,TLC chamber, aluminium foil and capillary tube.
- UV light at 254 nm.
- Water pump.

5.34 Extraction of Alkaloids from Chinese Herbs- *Coptis Chinensis*

- Alumina TLC plate, TLC chamber, aluminium foil and capillary tube.
- 250 mL Beaker.
- Büchner funnel and suction flask.
- Condenser.
- Distillation head, Liebig condenser, and receiver adapter.
- 100 mL Erlenmeyer flask.
- Filter paper and rubber adapter.
- 100 mL Graduated cylinder.
- Heating set-up.
- Iron stand and clamps.
- Melting point apparatus and melting tubes.

- Pasture pipette.
- 250 mL Round bottom flask.
- Rubber tubing.
- 150 mL Soxhlet extractor with a filter paper.
- Thermometer and adapter.
- UV light at 254 nm.
- Water pump.

5.35 Extraction of Caffeine from Teas

- 250 mL & 50 mL Beaker.
- Büchner funnel and suction flask.
- 250 mL Erlenmeyer flask.
- Filter paper and rubber adapter.
- 100 mL Graduated cylinder.
- Heating set-up.
- Iron ring, iron stand and clamps.
- Melting point apparatus and melting tubes.
- Petri dish.
- Rota-evaporator.
- Rubber tubing.
- 250 mL Separatory flask.
- Water pump.

6.2 List of Chemicals① for Each Experiment

5.1 Purification of Liquids by Distillation

Aqueous ethanol.

5.2 Purification of Organic Compounds by Extraction

Aqueous acetic acid, diethyl ether, 0.2000 mol · L^{-1} standardized NaOH solution, 0.1% phenolphthalein in 95% ethanol (indicator).

5.3 Purification of Solids by Recrystallization

N-Phenyl acetamide, activated carbon, water.

---

① All chemicals should be at least in the Reagent grade in the experiments listed in this book.

5.4 Purification of *Cyclo*hexane

Phenol contaminated *cyclo*hexane, 1% $FeCl_3$ aqueous solution, 5% NaOH solution, water, anhydrous $CaCl_2$.

5.5 Preparation of Bromoethane

NaBr, concentrated H_2SO_4, 95% EtOH, H_2O.

5.6 Synthesis of Ethyl Acetate

Ethanol, acetic acid, Conc. H_2SO_4, saturated sodium bicarbonate solution, saturated calcium chloride solution, anhydrous magnesium sulfate, brine.

5.7 Synthesis of Benzoin

Benzaldehyde, thiamine (vitamin B_1), ethanol, distilled water, activated carbon, sodium hydroxide, 20% ethyl acetate in petroleum ether.

5.8 Synthesis of Cinnamic Acid by the Perkin Reaction

Benzaldehyde, acetic anhydride, potassium carbonate, conc. hydrochloric acid, water, 10% sodium hydroxide solution, activated carbon.

5.9 Synthesis of Cinnamic Acid by the Knoevenagel Condensation

Benzaldehyde, malonic acid, potassium carbonate, 2 mol · L^{-1} hydrochloric acid, pyridine, piperidine.

5.10 Synthesis of Methyl Orange

Sulfanilic acid, *N*, *N*-dimethylaniline, conc. HCl, sodium nitrite, acetic acid, 5% and 10% NaOH aqueous solution, sodium chloride, brine, ethanol, diethyl ethe.

5.11 Synthesis of Ethyl Benzoate

Ethanol, benzoic acid, conc. H_2SO_4, *cyclo*hexane, 10% sodium carbonate, diethyl ether, anhydrous calcium chloride, 25% EtOAc/petroleum ether.

5.12 The Diels-Alder Reaction

Xylenes, anthracene, maleic anhydride, dichloromethane, 10% EtOAc in petroleum ether.

5.13 Synthesis of Acetanilide

Aniline, glacial acetic acid, zinc powder, water, 10% methanol/

dichloromethane.

5.14 Bromination of Acetanilide

Potassium bromate, 48% hydrobromic acid in water, acetanilide, glacial acetic acid, 95% ethanol, 10% sodium bisulfite.

5.15 Deprotection of Acetyl Protecting Group from *p*-Bromoacetanilide

Ethanol, *p*-bromoacetanilide, potassium hydroxide, water, ethyl acetate, petroleum ether, methanol, dichloromethane.

5.16 Synthesis of Aspirin

Salicylic acid, acetic anhydride, conc. sulfuric acid, water, saturated sodium carbonate solution, conc. hydrochloric acid, ethyl acetate, petroleum ether, methanol, dichloromethane, ethanol.

5.17 Synthesis of Ferrocene via the Phase Transfer Catalysis

Tetrahydrofuran, *cyclo*penta-1,3-diene (freshly cracked by the instructor), 18-crown-6, potassium hydroxide, ferrous chloride tetrahydrate, conc. hydrochloric acid, ice, water, hexane, ethyl acetate, petroleum ether.

5.18 Synthesis of 2-Nitro-1,3-benzenediol

1,3-Dihydroxybenzene, conc. H_2SO_4, 65%～68% nitric acid, urea, ethanol, water, 20% ethyl acetate in petroleum ether, ice.

5.19 Preparation of the Urea-Formaldehyde Resin

Formaldehyde solution (～37%), concentrated ammonia aqueous solution, urea, 1% sodium hydroxide solution, ammonium chloride.

5.20 Preparation of *Cyclo*hexene

*Cyclo*hexanol, conc. phosphoric acid, sodium chloride, 5% sodium carbonate solution, brine, anhydrous calcium chloride, boiling chips, 1% bromine in CCl_4.

5.21 Synthesis of 2-Chloro-2-methylpropane

Conc. HCl, *tert*-butanol, 5% sodium bicarbonate aqueous solution, sodium bicarbonate, anhydrous calcium chloride.

5.22 Synthesis of Dibutyl Ether

Boiling chips, butan-1-ol, conc. H_2SO_4, 50% H_2SO_4 aqueous solution, anhydrous calcium chloride.

5.23 Synthesis of Butyl Phenyl Ether

Phenol, sodium hydroxide, 1-bromobutane, 6% tetrabutylammonium bromide, 10% sodium hydroxide solution, brine, anhydrous sodium sulfate, water, ethyl acetate, petroleum ether, dichloromethane.

5.24 Synthesis of Hexane-1,6-dionic Acid

*Cyclo*hexene, potassium permanganate, water, conc. hydrochloric acid, methanol, 1% sodium hydroxide solution.

5.25 Preparation of Furfuryl Alcohol and 2-Furoic Acid by the Cannizzaro Reaction

43% sodium hydroxide, furan-2-carbaldehyde, diethyl ether, anhydrous magnesium sulfate, conc. hydrochloric acid, 1 : 1 EtOAc: petroleum ether, 1 : 9 methanol: dichloromethane, water.

5.26 A Solvent Free Cannizzaro Reaction

Potassium hydroxide, 2-chlorobenzaldehyde, ethanol, conc. hydrochloric acid, 50% ethyl acetate in petroleum ether, 5% methanol in dichloromethane, water.

5.27 Reduction of 3-Nitroacetophenone using Sodium Borohydride

Sodium borohydride, 3-nitroacetophenone, ethanol, toluene, ethyl acetate, anhydrous sodium sulfate, 1 : 1 ethyl acetate: petroleum ether.

5.28 Reduction of 3-Nitroacetophenone Using Sn/HCl

Granular Sn, 3-nitroacetophenone, conc. hydrochloric acid, 5% methanol in dichloromethane, dichloromethane, 30% sodium hydroxide aqueous solution.

5.29 Synthesis of 4-Vinylbenzoic Acid via the Wittig Reaction

4-Bromomethylbenzoic acid, triphenylphosphine, acetone, diethyl ether, 37% formaldehyde aqueous solution, sodium hydroxide, water, ethanol, conc. HCl, dichloromethane, 50% ethyl acetate in dichloromethane.

5.30 Preparation of 1,5-Diphenylpenta-1,4-dien-3-one Using the Aldol Condensation

Benzaldehyde, sodium hydroxide, acetone, ethanol, ethyl acetate, 10% ethyl acetate in petroleum ether.

5.31 Preparation of 2-Methylbutan-2-ol via the Grignard Reagent

Granular magnesium or magnesium strip, ethyl bromide, iodine, anhydrous acetone, anhydrous diethyl ether, 20% sulfuric acid aqueous solution, anhydrous calcium chloride, anhydrous sodium carbonate.

5.32 Synthesis of Dimedone via the Robinson Annulation

Sodium ethoxide, diethyl malonate, 4-methylpent-3-en-2-one, anhydrous ethanol, 60～90 ℃ petroleum ether, diethyl ether, potassium hydroxide, conc. hydrochloric acid.

5.33 Preparation of Phenacetin from *p*-Aminophenol via a Two-Step Synthesis

Acetic anhydride, *p*-aminophenol, sodium ethoxide, iodoethane, 20% Petroleum ether in diethyl ether.

5.34 Extraction of Alkaloids from Chinese Herbs- *Coptis Chinensis*

Coptis chinensis, ethanol, 1% acetic acid aqueous solution, 6 M hydrochloric acid, 10% methanol in chloroform.

5.35 Extraction of Caffeine from Teas

Tea, water, sodium carbonate, dichloromethane, 95% ethanol, Celite®

6.3 Appendix Tables

Table 6.1 Table of physical properties for commonly used solvents

Solvent	Boiling point (℃)	Density ($g \cdot mL^{-1}$)	Miscible with H_2O
Acetone	56.5	0.792	√
Butan-1-ol	74.1	0.810	slightly
Butan-2-ol	74.1	0.806	slightly
Butan-2-one	72.1	0.800	soluble
tert-Butanol	82.2	0.78	√
CH_3CN	81.65	0.79	√
CH_2Cl_2	40	1.326	×
$CHCl_3$	61.2	1.48	×
CCl_4	153.8	1.60	×

(Continued)

Solvent	Boiling point (℃)	Density (g · mL^{-1})	Miscible with H_2O
*Cyclo*hexane	80.7	0.78	×
DMF	153	0.94	√
DMSO	189	1.10	√
Et_2O	35	0.713	×
EtOAc	77	0.902	slightly
EtOH	78.5	0.789	√
Hexane	69	0.660	×
HOAc	117.9	1.05	√
MeOH	65	0.792	√
NMP	204	1.03	√
Pentane	36	0.626	×
PhH	80.1	0.88	×
PhMe	111	0.866	×
Propan-1-ol	97	0.803	√
Propan-2-ol	82.5	0.785	√
THF	66	0.89	√
o-Xylene	144	0.897	×
m-Xylene	139.1	0.868	×
p-Xylene	138.4	0.861	×

Table 6.2 Table of different grades of chemicals

Grade	Definition
ACS	Purity of the chemical is same or higher than the stand purity regulated by the *American Chemical Society*.
Reagent	High purity equals to the ACS grade and suitable for laboratory use and analytical applications and it is the highest quality commercially available for this chemical.
GR	Short for guaranteed reagent. Purity of the chemical satisfies the *American Chemical Society* requirements and chemicals of this grade are generally good for analytical chemistry.
AR	It is the standard Mallinckrodt grade of chemicals for analysis. Purity of the chemical meets requirements on analytical reagents by the *American Chemical Society*. It is suitable for laboratory and general use.
OR	Organic reagents which are used for research purposes.

(Continued)

Grade	Definition
USP	Purity of the chemical is sufficient to meet the requirements of the *United States Pharmacopeia* and widely used for food, drug, or medicinal uses as well as for most laboratory purposes.
NF	Purity of the chemical satisfies requirements of the *National Formulary*.
Laboratory	Purity of the chemical is relatively high but the actual amount/level of impurities is unknown. It is pure enough for education applications but not for food, drug or medicinal use.
Purified	Also referred to pure or practical grade, indicates chemicals of good quality which matches no official standard. It is used in most cases for education application but not for food, drug or medicinal use.
CP	Short for chemically pure. The product of this grade is fine for use in general applications.
Technical	Purity of the chemical is of good quality for commercial and industrial purposes but not for food, drug or medicinal use.
HPLC/Spectro	Purity of the chemical meets the requirements of standard chromatography in HPLC and spectrophotometry procedures.
Omni HPLC	The quality of the product satisfies the ACS requirements for use in HPLC and Ultraviolet spectrophotometry.
ChromAR	Solvents that satisfy ACS specifications and good for liquid chromatography and UV-spectrophotometry.
SpectraAR	Solvents good for UV-spectrophotometric applications.

Table 6.3 Table of several commonly used chemical sources

Vendors	Websites
Acros Organics	https://www.acros.com
Aladdin Chemical	http://www.aladdin-e.net
Alfa Aesar	https://www.alfa.com/en/
Energy Chemical	https://www.energy-chemical.com/front/index.htm
J&K	https://www.jkchemical.com/CH/Index.html
Macklin	http://www.macklin.cn
Sigma-Aldrich	https://www.sigmaaldrich.com
Tansoole	http://www.tansoole.com
TCI Chemicals	https://www.tcichemicals.com

Table 6.4 Referenced chemical shifts of some commonly used deuterated solvents

Solvent	δ (^{1}H, ppm) (*multiplicity*)	δ (^{13}C, ppm) (*multiplicity*)
$CDCl_3$	7.28 (1)	77.2 (3)
CD_3COCD_3	2.04 (5)	29.8 (7), 206.3 (1)
C_6D_6	7.26 (1)	128.4 (3)
CD_3CN	1.93 (5)	1.4 (7), 118.7 (7)
CD_2Cl_2	5.32 (3)	53.5 (5)
CD_3SOCD_3	2.50 (5)	39.5 (7)
CD_3OD	3.35 (5), 4.78 (1)	49.2 (7)
D_2O	4.80 (1)	—
CD_3COOD	2.03 (5), 11.53 (1)	20.0 (7), 178.4 (1)

Table 6.5 Table for the azeotropic boiling points of various azeotropes

Azeotrope components	Azeotropic boiling point (℃)	Azeotrope components	Azeotropic boiling point (℃)
95.5%* EtOH in H_2O	78.1	58.3% *cyclo*hexanol in H_2O	97.8
71.7% PrOH in H_2O	87.7	79.8% PhMe in H_2O	79.8
87.9% ^{i}PrOH in H_2O	80.4	98.7% Et_2O in H_2O	34.2
55.5% BuOH in H_2O	92.7	95% THF in H_2O	65
67.9% ^{sec}BuOH in H_2O	88.5	59.5% PhOMe in H_2O	59.5
70.0% ^{i}BuOH in H_2O	90.0	83.7% MeCN in H_2O	76.5
88.3% ^{t}BuOH in H_2O	79.9	EtOH-*cyclo*hexane—H_2O	62.1
97% $CHCl_3$ in H_2O	53.3	EtOH-hexane—H_2O	56.0
95.9% CCl_4 in H_2O	66.8	EtOH-PhH—H_2O	64.9
99.6% CH_2Cl_2 in H_2O	38.8	EtOH-$CHCl_3$—H_2O	55.5
91.9% EtOAc in H_2O	70.4	EtOH-PhMe—H_2O	74.4
57% PyH in H_2O	92.6	EtOH-MeCN—H_2O	72.9
91.1% PhH in H_2O	69.3	Bu_2O in H_2O	92.9
58.4% *cyclo*hexene in H_2O	70.8	2-methylbutan-2-ol in H_2O	87.4
58.3% *cyclo*hexane in H_2O	69.5	*Cyclo*hexene in *cyclo*hexanol	64.9

* This is the percentage by weight.

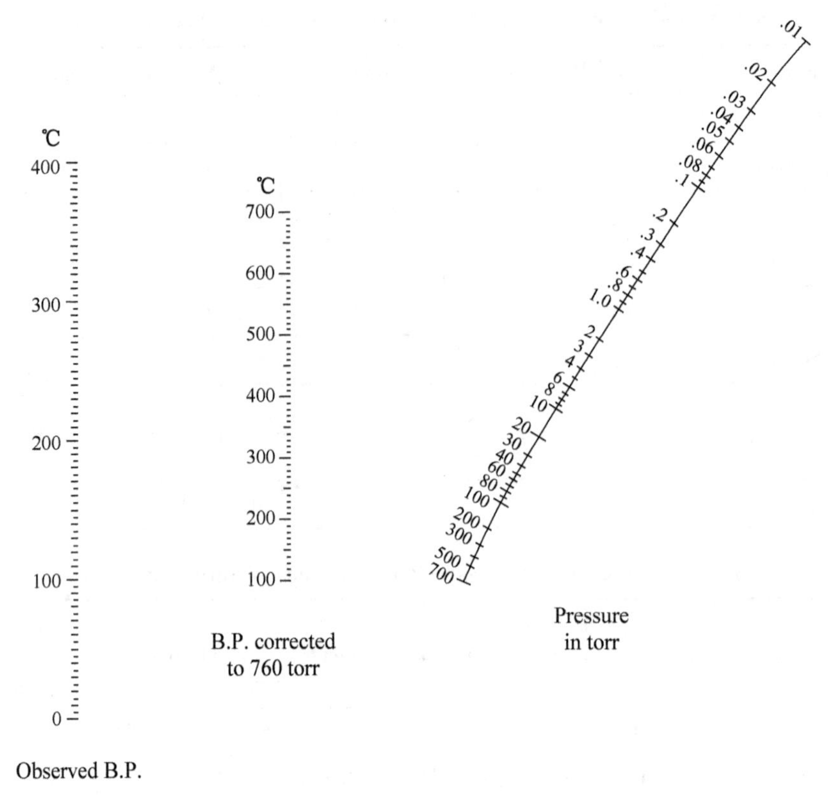

Figure 6. 1 A nomograph

6.4 Abbreviations

AcOH	acetic acid
ACS	American Chemical Society
amu	atomic mass unit
AR	analytical reagent
B. P.	boiling point
br	broad
cmpd	compounds
conc.	concentrated
COSHH	Control of Substance Hazardous to Health
CP	chemically pure
DCM	dichloromethane
DMF	*N*, *N*-dimethylformaldehyde
DMS	documentation of molecular spectroscopy
DMSO	dimethyl sulfoxide
EI	electron impact

ESI	electrospray ionization
EtOAc	ethyl acetate
eV	electron voltage
FID	free induction decay
FT	Fourier transform
GC	gas chromatography
GHS	Globally Harmonized System of Classification and Labelling of Chemicals
GR	guaranteed reagent
h	hour
HPLC	high performance liquid chromatography
HRMS	high-resolution mass spectroscopy
IR	infrared
LC	liquid chromatography
m	medium
m/z	mass-to-charge ratio
MALDI	matrix-assisted laser desorption ionization
Me	methyl
MED	medium
min	minute
mp	melting point
MS	mass spectroscopy
MSDS	material safety data sheet
MW	molecular weight
NF	national formulary
NMR	nuclear magnetic resonance
NMP	*N*-methylpyrrolidone
OR	organic reagents
P. T.	proton transfer
Ph	phenyl
PhH	benzene
PhMe	toluene
ppm	part per million
PTFE	polytetrafluoroethylene
RBF	round bottle flask
R_f	retention factor
rota	rotatory
s	strong

SDS	safety data sheet
sep	separatory
THF	tetrahydrofuran
TLC	thin layer chromatography
TMS	tetramethylsilane
TOF	time-of-flight
UV	ultraviolet
Vis	visible
w	weak
γ	magnetic moment
δ	chemical shift
ν	frequency